U0100058

零失敗秘方系列

煮出香濃伴飯餸

Rich and delectable home dishes

編者話

Preface

肉香、海鮮濃、咖喱惹味……每一口,都令人吃上癮。

香濃的菜式不是單靠醬料就能成事,還要考慮材料如何選擇、配料如何搭配、料頭調味如何適當加添……惹味的餸菜不難出現在你家的飯桌上。

任何食材,都可變成惹味菜式的主角,簡單如一個雞蛋、一盤蜆仔等,配上自家調配的醬汁或天然香料,加上烹煮的巧手及心意,在家炮製鮮香濃味的餸菜又有何難度?

今晚就來變身大廚,為你的惹味晚餐揭開序幕吧!

Rich and tangy meat dishes, sweet and flavoursome seafood, spicy and fragrant curry... Just one bite is enough to get you addicted.

A rich and flavoursome dish isn't always the sole effort of a good sauce. You must formulate the right combination of key ingredients, side ingredients, aromatics and seasoning. Dishes that pack oomph and big flavours aren't that difficult to turn out from your kitchen if you know the tricks.

Any ingredient can be the hero of a flavoursome dish. Even an egg or a bowl of clams that seem unspectacular on their own, can use a makeover to be a shining star, as long as you pair them with the right homemade sauces or natural spices, and cook them with the right techniques and with all your heart. You're never more than a corner away from a smorgasbord bursting with bold flavours.

So, get in the kitchen now and start building layers of flavours for your loved ones.

目錄

···

CONTENTS

香辣 Spicy

金不換辣椒膏炒青口 / 72
Stir-fried Mussels with Thai Basil and
Chilli Paste

口水雞 / 74
Steamed Chicken Dressed in
Sichuan Peppercorn Chilli Oil

咖喱三文魚頭粉皮煲 / 78
Simmered Salmon Head and Vermicelli
Sheet with Curry Sauce in Clay Pot

泡菜年糕牛仔骨 / 81
Beef Short Ribs with Kimchi and
Rice Cakes

蒜香豆豉辣椒醬香草炒蜆 / 84
Stir-fried Clams with Mint, Black Bean
and Garlic Chilli Sauce

黑椒醬爆蟶子皇 / 88
Stir-fried King Razor Clams with
Black Pepper Sauce

麻辣炮椒雞鍋 / 90
Spicy Bullet Chilli Chicken Casserole

XO 醬翠玉瓜炒魚塊 / 92
Stir-fried Fish Fillet with Zucchini in
XO Sauce

椰香咖喱蟹 / 96
Curry Crab with Coconut Milk

杳茅辣椒炒東風螺 / 98
Stir-fried Spiral Babylon with
Lemongrass and Chillies

酥香 Crispy

魚腸涼瓜煎蛋 / 100
Egg Omelette with Fish Intestines and
Bitter Melon

香酥芝麻青芥辣雞中翼 / 104
Deep-fried Chicken Wings with
Sesame Seeds and Wasabi

脆炸門鱔肉 / 107
Deep-fried Conger-pike Eel

南乳碎炸雞 / 110
Deep-fried Chicken with
Fermented Tarocurd

美味秘笈：
為菜式加添惹味元素
Tricks to bold flavours:
adding an oomph to your dishes

　　想炮製惹味香濃的餸菜，一般人想到的是依賴醬料，拌一勺香醬令菜式生色不少。其實，利用其他天然材料或香料，任何簡單的菜式都成為「超級」好餸，來吧！一起為入廚增長知識！

- 大量使用料頭，如薑、葱、蒜、乾葱頭、京葱等，熱油爆炒香氣四溢，再配上肉類或主料同煮，提升不少香味。
- 選用自家調配的醬料，鹹味及油分可自行控制，可變出獨一無二的醬汁來，令煮餸事半功倍。
- 多選用天然香料及醬汁，如芫茜、黑椒、羅望子、香茅、辣椒、花椒、金不換等，為餸菜帶來意想不到的驚喜。
- 啤酒、南乳及腐乳等發酵食材，其濃郁的味道，是令餸菜惹味的另一元素。

- 橙、檸檬、柑桔、菠蘿、玫瑰花蕾等散發清香氣味，加添菜式內烹調，為材料增添幽香的味道，令人難忘！
- 嗜辣的話，咖喱、紅辣椒、黑椒、川椒必定不好缺少，用油爆炒後煮成醬汁更加美味，辣度自行調校，伴白飯吃一流！
- 即使購買冰鮮食材，處理好解凍、去除冷藏味及醃味等步驟，同樣嘗到香濃的美味菜式。

When people think of richly flavoured dishes, most of them would look for pre-made sauces in a bottle. Just a spoonful of sauce would shift the flavour profile and give a dish a lot more character. In fact, you can turn any simple home-style dish into a stellar creation with natural condiments and spices. Let's learn and be a better cook.

- You may use aromatics, such as ginger, spring onion, garlic, shallot or Peking scallions, generously to add aromas and flavours. Just fry them in hot oil until fragrant before adding meat or other key ingredients.
- You may make your own sauces so that you have control over the saltiness and the amount of oil used, and customize the taste according to your needs. With your homemade sauces in the fridge, you can easily and conveniently season your dishes.
- Natural spices and condiments are a great source of flavours. Try coriander, black pepper, tamarind, lemongrass, chilli, Sichuan peppercorns or Thai basil. They would bring pleasant surprise to your palate.
- Fermented food items such as beer, fermented tarocurd and fermented tofu have a complex taste profile themselves, with a characteristic richness. They could add zing to the food instantly.

- Food items with essential oil, such as orange, lemon, kumquats, pineapple, and rose buds carry a refreshing and light fragrance. They also add a fruitiness or floral nose to your dish, making it unforgettable.
- Those who want more heat in their food aren't strangers to piquant ingredients like curry, red chillies, black pepper and Sichuan peppercorns. Just fry them in hot oil until fragrant and make a sauce with them. Adjust the amount used for your preferred spiciness. They make great dishes to go with rice.
- You don't always need to use fresh ingredients for heavily seasoned dishes. Frozen or chilled ingredients work equally well if you thaw them correctly and marinate them to remove the unpleasant smell of frozen food.

頭抽煎焗大鱔

Fried Eel with Premium Soy Sauce

◎ 材料（4 人份量）

白鱔 12 兩
陳皮 1/8 個
蒜茸 2 湯匙
葱粒 1.5 湯匙

◎ 醃料

鹽 2/3 茶匙
胡椒粉少許
生粉半湯匙（後下）

◎ 調味料（拌勻）

水 1 湯匙
頭抽 2 湯匙
糖 1 茶匙

◎ 做法

1. 白鱔去掉潺液，切成半吋厚鱔件
   （魚販可代勞），洗淨，抹乾。在
   鱔皮上往肉處切成 V 型，下醃料
   醃 10 分鐘（煎時才拌入生粉）。

2. 陳皮用水浸軟，刮去瓤，切絲。

3. 易潔鑊下少許油，放入鱔件煎至微
   黃色，反轉再煎，加入陳皮絲及蒜
   茸，加蓋焗片刻。

4. 灒入調味料，煮至汁液乾透，最後
   灑入葱粒，上碟享用。

◎ Ingredients (Serves 4)

450 g white eel
1/8 dried tangerine peel
2 tbsp finely chopped garlic
1.5 tbsp diced spring onion

◎ Marinade

2/3 tsp salt
ground white pepper
1/2 tbsp caltrop starch (added at last)

◎ Seasoning (mixed well)

1 tbsp water
2 tbsp premium soy sauce
1 tsp sugar

◎ Method

1. Remove slime from white eel and
   cut into slices of 1/2 inch thick (or ask
   the fish monger for help). Rinse and
   wipe dry. Slit V shapes at the skin and
   marinate for 10 minutes (add caltrop
   starch right before frying).

2. Soak dried tangerine peel until soft.
   Scrape off the pith and cut into
   shreds.

3. Heat a little oil in a non-sticky frying
   pan. Fry the eel slices until slightly
   golden. Fry the other side. Put in
   shredded dried tangerine peel and
   garlic. Cover the lid and cook for a
   while.

4. Pour in the seasoning. Cook until the
   sauce is dry. Add diced spring onion
   and serve.

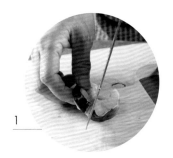

1

2

◎ 零失敗技巧 ◎
Successful Cooking Skills

為何在鱔皮上切成 V 型？

使鱔皮斷開，煎時不會太蜷曲，賣相更佳。

Why slit V shapes at the eel skin?

It breaks open the skin so that it would not curl up during frying and it appears good.

用易潔鑊煎白鱔有何好處？

由於白鱔油分多，可省卻用油量，而且減低白鱔之油膩感。

What are the benefits of frying eel with non-sticky pan?

Since white eel is rich in oil, it can save a lot of oil and also reduce the greasiness of the eel.

可用普通生抽代替頭抽嗎？

可以，但味道卻比頭抽略遜。

Can premium soy sauce replaces by normal light soy sauce?

Yes but the flavor reduces.

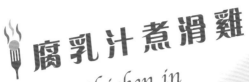 腐乳汁煮滑雞

Chicken in Fermented Tofu Sauce

◎ 材料 （4 人份量）
光雞半隻（斬件）
腐乳 3 塊
乾雞蹤菌半兩
薑 6 片
蒜肉 4 粒
紅椒絲少許
紹酒 1 湯匙

◎ 醃料
胡椒粉少許
粟粉 1 茶匙

◎ 調味料
糖半茶匙
生抽 2 茶匙
水 1 杯

腐乳汁煮滑雞

◎ 做法

1. 雞蹤菌用水浸軟，洗淨。

2. 腐乳加水 1 湯匙拌成腐乳汁。

3. 雞塊洗淨，隔去水分，加入醃料拌勻。

4. 燒熱鑊，下油 1 湯匙，下薑片及蒜肉炒香，加入腐乳汁炒勻，下雞塊及雞蹤菌炒片刻，灒酒，拌入調味料用中火煮 15 分鐘，盛起，以紅椒絲裝飾即成。

◎ Ingredients (Serves 4)
1/2 chicken (chopped into pieces)
3 cubes fermented tofu
19 g dried termite mushrooms
6 slices ginger
4 cloves skinned garlic
shredded red chilli
1 tbsp Shaoxing wine

◎ Marinade
ground white pepper
1 tsp cornflour

◎ Seasoning
1/2 tsp sugar
2 tsp light soy sauce
1 cup water

◎ Method

1. Soak the termite mushrooms in water until soft. Rinse well.

2. Mix the fermented tofu with 1 tbsp of water as sauce.

3. Rinse the chicken pieces. Drain. Mix well with the marinade.

4. Heat a wok. Pour in 1 tbsp of oil. Stir-fry the ginger and garlic until fragrant. Add the fermented tofu sauce. Stir-fry and mix well. Put in the chicken and mushrooms and stir-fry for a while. Sprinkle with the wine. Add the seasoning and cook over medium heat for 15 minutes. Garnish with the red chilli and serve.

◎ 零失敗技巧 ◎
Successful Cooking Skills

有何烹飪要點？

必須留意腐乳之鹹味，若味道太鹹，可加入少許糖及去掉調味料之生抽份量。

Any cooking tips?

If the fermented tofu is too salty, add a little of sugar and reduce the amount of light soy sauce for seasoning.

如何挑選腐乳？

購買本地醬園出產的腐乳為佳，味道特別香滑，鹹味適中。若見腐乳質地硬實，不宜購買。

How to choose fermented tofu?

The best choice is those from local sauce factories. Smelling great, the fermented tofu is very smooth and moderately salty. If it looks rather firm, do not buy it.

哪裡有售乾雞蹤菌？

日式大型超市、有機食品店及菇菌雜貨店均有出售。

Where to buy dried termite mushrooms?

They are available at large Japanese supermarkets, organic food stores and grocery shops selling mushrooms.

乾煎大蝦伴鹹檸檬醬

Seared Prawns with Salted Lemon Dip

材料 （3 至 4 人份量）

新鮮大蝦 6 隻（約 12 兩）
乾粟粉適量

醃料

胡椒粉少許

蝦頭調味料

鹽、胡椒粉各少許

蘸汁調味料

鹹檸檬 1/4 個（見 p.16）
糖 2 茶匙
檸檬汁 2 湯匙
凍開水 2 湯匙

做法

1. 鹹檸檬去核，切碎，加入調味料，放於攪拌機打成茸作蘸汁，備用。

2. 大蝦剝去蝦頭（留用），挑腸，洗淨，去蝦殼、留蝦尾。由蝦背剒雙飛，拍扁，下醃料拌勻醃片刻。

3. 蝦頭洗淨，抹乾水分，放入油鑊煎至金黃全熟，盛起，下調味料拌勻。

4. 全隻蝦蘸上乾粟粉，炸至金黃全熟，隔油，排上碟，伴蝦頭及蘸汁食用。

Ingredients (Serves 3-4)

6 fresh large prawns (450 g)
cornflour

Marinade

ground white pepper

Seasoning for prawn heads

salt
ground white pepper

Dipping sauce

1/4 salted lemon (see p.16)
2 tsp sugar
2 tbsp lemon juice
2 tbsp cold drinking water

Method

1. To make the dipping sauce, seed the salted lemon and finely chop it. Add the remaining ingredients for the dip. Puree in a blender. Set aside.

2. Remove the heads of all prawns and keep them for later use. Devein the prawns and rinse well. Shell them but keep the tail intact. Make a cut along the back of each prawn to butterfly it. Pat gently with the flat side of a knife. Add marinade and stir well.

3. Rinse the prawn heads well. Wipe dry. Fry in oil until golden and done. Drain. Add seasoning and stir well.

4. Coat the prawns lightly in cornflour. Deep fry in hot oil until golden and done. Drain. Arrange on a serving plate. Place the prawn heads on dish and serve the dipping sauce on the side.

* 鹹檸檬製法
Making salted lemon

乾煎大蝦伴鹹檸檬醬

◎ 材料
檸檬 10 個
粗鹽 2 磅（或適量）

◎ 做法
1. 檸檬洗淨外皮，風乾一晚至乾透。
2. 玻璃瓶內鋪入一層粗鹽，排入檸檬及粗鹽（至排入全部檸檬），最後面層鋪滿粗鹽，以免檸檬發霉。

◎ Ingredients
10 lemons
900 g coarse salt (enough to cover all lemons)

◎ Method
1. Rinse the lemons and leave them to dry overnight.
2. In a sterilized glass container, pour in a layer of coarse salt. Arrange lemons on top of the salt. Top with another layer of salt. Repeat this step until all lemons are used. Top generously with a last layer of salt to cover all lemon well. The lemon might turn mouldy if exposed to air.

◎ 零失敗技巧 ◎
Successful Cooking Skills

醃製檸檬有何成功要訣？

宜洗淨檸檬外皮，風乾一晚，待外皮呈少許皺紋，再用粗鹽長時間醃製。

What is the technique for marinating the salted lemon?

Rinse the lemons well and leave them to dry overnight. Their peels would turn slightly wrinkly. Then pickle them in salt for a long time.

蝦剝雙飛後，為何要輕輕拍扁？

用刀輕拍蝦隻，令蝦筋鬆斷，乾煎後令肉質收縮，外型美觀。

Why do you pat the prawns lightly after butterflying them?

Patting them with the flat side of a knife helps break the tendon on the prawns. The flesh will shrink after it is cooked and gives a better presentation.

蘸汁調味料為何加入檸檬汁？

令醬汁多一分層次，提升檸檬的鮮味，令檸檬蘸醬更香濃。

Why do you add lemon juice to the dip?

It adds another dimension to the dip by accentuating the zesty lemon flavour. The dip will taste stronger and richer this way.

香茜豆醬雞扒

Fried Chicken Steak with Coriander in Bean Sauce

◎ 材料 （4 人份量）
雞扒 2 件（約 12 兩）
芫茜 2 棵
普寧豆醬 2 湯匙
芝麻醬 1 湯匙

◎ 醃料
生抽 3/4 湯匙
老抽 1 茶匙
薑汁酒 1 湯匙
胡椒粉少許
生粉半湯匙

◎ 獻汁
水半杯（125 毫升）
糖半茶匙
胡椒粉及麻油各少許
生粉半茶匙

◎ 做法

1. 雞扒解凍，洗淨，抹乾水分，加入醃料拌勻約 15 分鐘。

2. 熱鑊下油，放入雞扒煎熟，瀝乾油分，切件。

3. 芫茜洗淨，切短度，分開芫茜莖及芫茜葉，備用。

4. 熱鑊下油，下芫茜莖炒香，傾入獻汁、普寧豆醬及芝麻醬煮滾，放入雞件拌勻，上碟，以芫茜葉裝飾。

Ingredients (Serves 4)

2 chicken steaks (about 450 g)
2 stalks coriander
2 tbsp Puning bean sauce
1 tbsp sesame sauce

Marinade

3/4 tbsp light soy sauce
1 tsp dark soy sauce
1 tbsp ginger juice wine
ground white pepper
1/2 tbsp caltrop starch

Thickening glaze

1/2 cup water (125 ml)
1/2 tsp sugar
ground white pepper
sesame oil
1/2 tsp caltrop starch

Method

1. Defrost the chicken steak. Rinse and wipe it dry. Mix with the marinade and rest for about 15 minutes.

2. Add oil in a heated wok. Fry the chicken until fully done. Drain and cut into pieces.

3. Rinse the coriander. Cut into short sections. Separate the stems from the leaves. Set aside.

4. Add oil in the heated wok. Stir-fry the coriander stems until fragrant. Pour in the thickening glaze, Puning bean sauce and sesame sauce. Bring to the boil. Put in the chicken and mix well. Arrange on the plate. Decorate with the coriander leaves. Serve.

零失敗技巧
Successful Cooking Skills

如何令雞扒更入味？
於雞扒的厚肉部位用刀輕剁數下，令醃料快速入味！
How to make the chicken steak more flavourful?
Slightly make a few cuts on the thick part of the chicken to let the marinade infuse into the meat quickly.

雞扒切件後煎香，可以嗎？
建議煎透後才切件，能保持雞肉的肉汁，而且肉質不會乾硬。
Is it good to cut the chicken steak into pieces before frying?
To keep the meat juicy, soft, and moist, it is better to fry the whole steak first.

柱候牛筋腩

Beef Tendon and Brisket in Chu Hou Sauce

<table>
<tr><td>

◎ 材料 （6 人份量）

急凍牛肋條（或牛坑腩）600 克

急凍牛筋 400 克

蘿蔔 600 克

薑 2 片

乾葱頭及蒜肉各 3 粒（略拍）

陳皮 1 角（浸軟、刮瓤）

八角 3 粒

柱候醬 2.5 湯匙

紹酒 1 湯匙

◎ 調味料

生抽及老抽各 2 茶匙

冰糖碎 1 湯匙

鹽適量

</td><td>

◎ 做法

1. 牛肋條解凍，放入滾水煮 10 分鐘，盛起，過冷河。

2. 燒滾水 5 杯，下牛肋條及薑煮約 45 分鐘，待涼，切件，湯汁留用。

3. 牛筋飛水，切件；蘿蔔去皮，切件。

4. 燒熱油 1 湯匙，下乾葱頭及蒜肉爆香，加入牛肋條及牛筋炒透，下柱候醬炒至散發香味，灒酒，傾入牛肋條湯汁 2 杯、陳皮及八角，煮約 40 分鐘。

5. 最後加入蘿蔔及調味料再燜煮約 15 分鐘，待牛肋條軟腍即可享用。

</td></tr>
</table>

Ingredients (Serves 6)

600 g frozen beef rib finger (or beef brisket point end)

400 g frozen beef tendon

600 g radish

2 slices ginger

3 shallots (slightly crushed)

3 cloves skinned garlic (slightly crushed)

1 slice dried tangerine peel (soaked to soften; removed the pith)

3 star aniseed

2.5 tbsp Chu Hou sauce

1 tbsp Shaoxing wine

Seasoning

2 tsp light soy sauce

2 tsp dark soy sauce

1 tbsp crushed rock sugar

salt

Method

1. Defrost the beef rib finger. Cook in boiling water for about 10 minutes. Rinse with cold water.

2. Bring 5 cups of water to the boil. Cook the beef rib finger and ginger for about 45 minutes. Leave to cool down. Cut into pieces. Reserve the cooking sauce.

3. Scald the beef tendon. Cut into pieces. Peel the radish and cut into pieces.

4. Heat up 1 tbsp of oil. Stir-fry the shallot and garlic until scented. Add the beef rib finger and tendon. Stir-fry thoroughly. Put in the Chu Hou sauce and stir-fry until aromatic. Sprinkle with the wine. Pour in 2 cups of the cooking sauce. Add the dried tangerine peel and star aniseed. Cook for about 40 minutes.

5. Put in the radish and seasoning. Simmer for about 15 minutes until the beef rib finger is tender. Serve.

零失敗技巧
Successful Cooking Skills

牛肋條用薑片先煮，有何作用？

急凍牛肋條用薑片煮約 45 分鐘至半酥軟，除可去掉冷藏味，也可省掉燜煮的時間，加入醬料後燜煮，不容易黏底變焦。

What is the purpose of cooking beef rib finger with ginger beforehand?

By cooking the frozen beef rib finger with ginger for about 45 minutes until medium soft, the unpleasant flavour of frozen food can be removed and the simmering time also saved. It will not easily stick to the pot to get burnt after adding the sauce.

選購急凍的牛肋條還是新鮮的？

急凍的牛肋條徹底解凍後，飛水及過冷河，比新鮮的更容易軟腍。

Should I get frozen beef rib fingers, or fresh ones?

Frozen rib fingers tend to turn tender more quickly than fresh ones. Just thaw them completely. Blanch them in boiling water and rinse in cold water before use.

 雞絲粉皮

Mung Bean Sheets
with Shredded Chicken

◎ 材料 （3 人份量）

光雞半隻
鮮粉皮 5 張
指天椒 1 隻（去籽、切碎）
蒜頭 1 粒（剁茸）
白芝麻 1 湯匙
薑 2 片
蔥 2 條（取 1 條切粒）
紹酒半湯匙

◎ 醬汁

蒸雞原汁 1.5 湯匙
生抽 1 湯匙
鹽 1/4 茶匙
辣椒油、鎮江醋、糖及麻油各 2 茶匙
芝麻醬 5 茶匙

◎ 做法

1. 光雞洗淨，抹乾水分，將紹酒及鹽 3/4 茶匙抹勻雞身兩面，待 15 分鐘。

2. 將雞、薑片及蔥 1 條放於碟上，隔水蒸 20 分鐘，取出待涼，拆絲；蒸雞汁留用。

3. 粉皮切成粗條，放於碟內，鋪上雞絲，放入雪櫃冷藏約半小時。

4. 白芝麻用白鑊慢火炒香，備用。

5. 汁料拌勻，加入蒜茸、紅椒粒及蔥粒調和，澆在雞絲粉皮上，最後灑上白芝麻即成。

◎ Ingredients (Serves 3)

1/2 chicken
5 fresh mung bean sheets
1 bird's eye chilli (remove the seeds; crushed)
1 clove garlic (finely chopped)
1 tbsp white sesame seeds
2 slices ginger
2 sprigs spring onion (one diced)
1/2 tbsp Shaoxing wine

◎ Sauce

1.5 tbsp sauce from steaming chicken
1 tbsp light soy sauce
1/4 tsp salt
2 tsp chilli oil
2 tsp Zhenjiang vinegar
2 tsp sugar
2 tsp sesame oil
5 tsp sesame sauce

◎ Method

1. Rinse the chicken. Wipe it dry. Spread the Shaoxing wine and 3/4 tsp of salt on both sides of the chicken. Rest for 15 minutes.

2. Place the chicken, ginger and 1 sprig of spring onion on a plate. Steam over water for 20 minutes. Remove and allow it to cool down. Shred the chicken. Reserve the steaming sauce.

3. Cut the mung bean sheets into coarse strips. Place on a plate. Lay the shredded chicken on top. Refrigerate for about 1/2 hour.

4. Stir-fry the white sesame seeds in a wok without oil until scented. Set aside.

5. Mix the sauce well. Add the garlic, bird's eye chilli and diced spring onion. Mix well. Pour over the chicken and mung bean sheets. Sprinkle with the white sesame seeds and serve.

◎ 零失敗技巧 ◎
Successful Cooking Skills

有散裝鮮粉皮出售嗎？

售賣粉麵材料的店舖及南貨店有新鮮粉皮發售，每份約十張，毋須購買一大袋粉皮，數量太多用不完。

Are loose packing mung bean sheets available?

Fresh mung bean sheets are sold in raw noodle shops and groceries in portions of 10 sheets each. Do not buy a big pack as they would never be used up.

如何將醬汁調得如此惹味？

必須加入蒸雞原汁，令醬汁保留雞的原味，再滲有辣味、酸甜、麻油及芝麻香味，惹味程度直線上升！

How to make the flavour of the sauce such sensational?

The sauce gives an original chicken flavour by adding the sauce from steaming the chicken, compounded with the spicy, sweet and sour taste, and the fragrance of sesame oil and sesame seeds. It is absolutely fantastic!

烤原隻鮮魷魚
Grilled Whole Squids

◎ 材料 （6 人份量）

鮮魷魚（大）2 隻
西芹、甘筍、葱各 2 兩
洋葱烤肉醬半杯（見 p.26）

◎ 做法

1. 西芹、甘筍及葱全部洗淨，切絲。

2. 鮮魷魚抽出鬚頭及內臟，去掉軟骨，撕去外衣，洗淨，抹乾水分。

3. 用刀在鮮魷魚其中一面剝上橫紋，釀入蔬菜絲，排在墊上錫紙的焗盤，掃上洋葱烤肉醬。

4. 預熱焗爐 10 分鐘，放入鮮魷魚用 180℃焗 15 分鐘，掃上一層烤肉醬，再焗 10 分鐘，取出烤熟之魷魚，切塊供食。

◎ **Ingredients (Serves 6)**
2 large fresh squids
75 g celery
75 g carrot
75 g spring onion
1/2 cup onion barbecue sauce (see as below)

◎ **Method**

1. Rinse celery, carrot and spring onion. Finely shred them.

2. Dress the squids by pulling out their tentacles, head and innards. Remove the bone and peel off the outer purple skin. Rinse and wipe dry.

3. Make light straight incisions across the length on one side of the squids. Stuff them with vegetables. Arrange on a baking tray lined with aluminium foil. Brush onion barbecue sauce over the squids.

4. Preheat an oven for 10 minutes. Bake the squids at 180°C for 15 minutes. Brush on another layer of barbecue sauce. Bake for 10 more minutes. Cut into pieces and serve.

烤原隻鮮魷魚

* 洋葱烤肉醬
Onion Barbecue Sauce

◎ **材料**
洋葱半個（切絲）
蒜肉 8 粒（拍鬆）
黑胡椒碎 1 茶匙

◎ **調味汁（拌勻）**
蠔油、生抽、紹酒、麥芽糖各 2 湯匙
老抽 2 茶匙

◎ **做法**
燒熱鑊下油 3 湯匙，下洋葱、蒜肉及黑胡椒碎炒香，加入調味汁煮至濃稠，試味，待涼，入瓶儲存。

◎ **Ingredients**
1/2 onion (shredded)
8 cloves skinned garlic (crushed gently)
1 tsp ground black pepper

◎ **Seasoning (mixed well)**
2 tbsp oyster sauce
2 tbsp light soy sauce
2 tbsp Shaoxing wine
2 tbsp maltose
2 tsp dark soy sauce

◎ **Method**
Heat a wok and add 3 tbsp of oil. Stir fry onion, garlic and black pepper until fragrant. Add seasoning and cook until thick. Taste it. Leave it to cool. Store in sterilized bottles.

◎ 零失敗技巧 ◎
Successful Cooking Skills

洋蔥烤肉醬的鮮香取決於哪個步驟？
洋蔥及蒜肉炒至乾透，令醬汁香濃、美味。
How to make the onion barbecue sauce consistency richly aromatic?
The onion and garlic should be stir-fried until dry, for an intense flavour and dense consistency.

為何將蔬菜填入鮮魷內？
可將餘下的蔬菜廚餘填入魷魷內，令魷魚烤焗時不會太乾，保持濕潤。
Why do you stuff the squids with vegetables?
You can make good use of leftover vegetables this way. The vegetables keep the squids moist during the grilling process.

如何快速地撕掉魷魚外衣？
只需用刀在魷魚外衣上直剠一小段，撕開外衣往外一拉，整塊脫下。
How do you peel the purple skin of the squids quickly?
Just make a light incision on the outer skin of the squids. Then pull and peel the skin outward. It will come off in one piece.

雲耳啫啫滑雞煲
Sizzling Chicken with Cloud Ear in Clay Pot

◎ 材料（4 人份量）

冰鮮雞半隻
豬膶 2 兩
雲耳 1/3 兩
乾葱頭 6 粒
薑 6 片
葱白 4 段
三文鹹魚醬 2 茶匙（見 p.31）
紹酒 1 湯匙

◎ 醃料

生抽半湯匙
胡椒粉少許
粟粉 1 茶匙

◎ 調味料

老抽 2 茶匙
糖 1 茶匙
水 1 量杯

◎ 做法

1. 雲耳用水浸軟，去硬蒂，洗淨。
2. 冰鮮雞洗淨，斬件，下醃料拌勻，備用。
3. 豬膶洗淨，切厚片，加入紹酒 1 茶匙醃片刻。
4. 燒熱瓦煲下油 3 湯匙，下薑片、乾葱頭及葱白炒香，下三文鹹魚醬拌勻，加入雞塊炒透，潷酒，下雲耳及調味料加蓋煮 7 分鐘，最後下豬膶拌勻再煮 3 分鐘，聞到陣陣鹹魚的香味，瓦煲邊的汁液發出啫啫聲即成。

◎ Ingredients (Serves 4)

1/2 chilled chicken
75 g pork liver
13 g cloud ear fungus
6 cloves shallot
6 slices ginger
4 short segments of the white part of spring onion
2 tsps flaked salted salmon with dried scallops (see p.31)
1 tbsp Shaoxing wine

◎ Marinade

1/2 tbsp light soy sauce
ground white pepper
1 tsp cornflour

◎ Seasoning

2 tsp dark soy sauce
1 tsp sugar
1 measuring cup water

◎ Method

1. Soak the cloud ear fungus in water until soft. Cut off the hard stems. Rinse well.
2. Rinse the chicken and chop it into pieces. Add marinade. Mix well. Set aside.
3. Rinse the pork liver. Slice thickly. Add 1 tsp of Shaoxing wine. Mix well and leave it briefly.
4. Heat a clay pot and add 3 tbsp of oil. Put in the ginger, shallot and white part of spring onion. Add the flaked salted salmon with dried scallops. Stir well. Put in the chicken and sear well. Sizzle with wine. Add cloud ear fungus and seasoning. Cover the lid and cook for 7 minutes. Put in the pork liver at last and stir well. Cook for 3 minutes until you smell the fish and the sauce sizzles along the rim of the pot. Serve.

* 鹹三文魚製法
* Making salted salmon

1. 急凍三文魚 1 塊解凍，洗淨，抹乾水分。

2. 取瓦瓷或塑膠容器，先鋪一層粗海鹽，放入魚扒，最後蓋滿粗海鹽。

3. 加蓋，密封，冷藏於雪櫃的生果格層 2 至 3 星期。

1. Thaw 1 piece of frozen salmon. Rinse well and wipe dry.

2. Line a ceramic or plastic container with a layer of coarse sea salt. Put in the salmon and top with more sea salt until full.

3. Cover the lid and seal well. Store in the crisper compartment of your fridge for 2 to 3 weeks.

◎◎ 零失敗技巧 ◎◎
Successful Cooking Skills

為何稱為「啫啫煲」？

由於熱度十足，炒煮調味料及食材時發出「啫啫」之聲，因此而得名。

Why is it called "sizzling?"

Traditionally, this dish is served in a clay pot straight after heated strongly. The chicken is literally sizzling over the hot clay pot on the dining table.

豬膶炒煮 3 分鐘熟透嗎？

豬膶切厚片，進食時有嚼感。豬膶容易煮熟，煮 3 分鐘剛好熟透，嫩滑軟腍。

Why the pork liver is sliced thickly? Is it properly cooked in just 3 minutes?

It is sliced thickly for a more substantial mouthfeel and a lovely chew. It cooks very quickly and is cooked to perfection in 3 minutes. The liver should be soft and tender.

* 三文鹹魚醬製法

*Making flaked salted salmon with dried scallops

◎ 材料

鹹三文魚 6 兩

瑤柱半兩

薑茸 2 湯匙

蒜茸 1 湯匙

米酒 2 湯匙

粟米油 1.5 杯

◎ 做法

1. 瑤柱洗淨，用水浸軟，撕成瑤柱絲。

2. 鹹三文魚隔水蒸 10 分鐘，待冷，拆肉弄散。

3. 燒熱鑊下油半杯，下瑤柱絲炒香，加入薑茸炒香，下三文魚肉及蒜茸用慢火炒透及壓碎，潷酒，邊炒邊加入餘下的油（油必須蓋過所有材料），再炒片刻，盛起，待冷入瓶儲存。

◎ Ingredients

225 g salted salmon

19 g dried scallops

2 tbsp grated ginger

1 tbsp finely chopped garlic

2 tbsp rice wine

1.5 cups corn oil

◎ Method

1. Rinse the dried scallops. Soak them in water until soft. Tear them apart into fine shreds.

2. Steam the salted salmon for 10 minutes. Leave it to cool. Skin and de-bone it. Break it down into flakes.

3. Heat a wok and add 1/2 cup of oil. Put in the dried scallops and stir until fragrant. Add finely chopped ginger and stir fry until fragrant. Add flaked salmon and garlic. Stir fry until done and crush well. Sizzle with wine. Pour in the remaining oil while stirring continuously. There must be enough oil to cover all solid ingredients. Stir fry briefly. Set aside to let cool. Transfer into sterilized bottles.

海皇蝦湯脆米
Seafood and Prawn Broth with Crispy Rice

⊚ 材料（4 人份量）

斑肉 3 兩
中蝦 8 隻
鮮蟹肉 2 兩
白飯 1.5 碗
半生熟米飯 3 湯匙
西芹粒 2 湯匙
蝦湯 625 毫升

⊚ 蝦湯材料

大蝦頭、蝦殼共 12 兩
滾水 1 公升
香葉 3 片
白胡椒粒半茶匙

⊚ 醃料

鹽 1/4 茶匙
胡椒粉少許
生粉半茶匙

⊚ 做法

1. 蝦頭及蝦殼洗淨，瀝乾水分。燒熱油，下蝦頭及蝦殼煎香，注入滾水、香葉及胡椒粒，用慢火煮 1 小時，至蝦湯濃縮至 625 毫升，下半茶匙鹽調味。

2. 斑肉洗淨，抹乾水分，切件；中蝦去殼、去腸，洗淨，用醃料拌勻醃 10 分鐘。

3. 煲煮白飯期間，見飯水微滾，取出米飯成半生熟米飯，瀝乾水分。

4. 熱鑊下油，加入蝦球及魚肉炒熟，盛起。

5. 燒熱油，下半生熟米飯炸脆，瀝乾油分。

6. 西芹粒用滾水灼熟，盛起。

7. 鍋內放入白飯，下斑肉、蝦球、蟹肉及西芹粒，注入熱蝦湯，最後灑上脆米，伴湯享用。

⊚ 零失敗技巧 ⊚
Successful Cooking Skills

烹調蝦湯有何竅門？
必須煎香蝦頭及蝦殼，再用慢火熬煮至蝦湯濃郁，令湯味特別鮮香美味！
What are the tips of cooking prawn broth?
The prawn heads and prawn shells must be fried until fragrant first. Then cook them over low heat until the broth thickens.

如何自製鮮蟹肉？
購買花蟹或肉蟹（約重 12 兩），隔水蒸 14 分鐘，待涼，可拆出 3 至 4 兩蟹肉。
How to make crabmeat on my own?
Buy coral crab or mud crab (about 450 g) and steam for 14 minutes. Set aside to let cool and pick 113 g to 150 g of crabmeat.

Ingredients (Serves 4)

113 g grouper flesh
8 medium-sized prawns
75 g fresh crabmeat
1.5 bowls cooked rice
3 tbsp medium-cooked rice
2 tbsp diced celery
625 ml prawn broth

Ingredients of the prawn broth

450 g big prawn heads and shells
1 liter boiling water
3 bay leaves
1/2 tsp white peppercorns

Marinade

1/4 tsp salt
ground white pepper
1/2 tsp caltrop starch

Method

1. Rinse prawn heads and prawn shells. Drain. Heat oil in a wok and fry them until fragrant. Pour in boiling water, bay leaves and peppercorns. Cook over low heat for 1 hour until 625 ml of prawn broth remained. Season with 1/2 tsp of salt.

2. Rinse grouper flesh and wipe dry. Cut into pieces. Shell and devein medium-sized prawns. Rinse and marinate for 10 minutes.

3. Take up the rice when it is lightly boiled to give medium-cooked rice and drain.

4. Add oil into a hot wok. Put in shelled prawns and grouper flesh. Stir-fry until done and set aside.

5. Heat oil in wok. Deep-fry medium-cooked rice until crispy and drain.

6. Blanch diced celery in boiling water until done and drain.

7. Put cooked rice into a pot. Add grouper flesh, prawns, crabmeat and diced celery. Pour in hot prawn broth. Lastly sprinkle over the deep-fried rice and serve with the broth.

梅菜雞

Braised Chicken with Preserved Flowering Cabbage

◎ 材料 （4 人份量）
光雞 1 隻（約 2 斤）
梅菜 3 兩
蒜茸 1 湯匙
薑 2 片
葱 2 條
清水 1.5 杯

◎ 醃料
鹽半湯匙
老抽半湯匙
紹酒半湯匙

◎ 調味料
鹽 1/4 茶匙
糖半湯匙

◎ 生粉獻
水 1 湯匙
生粉 1 茶匙

梅
菜
雞

◎ 做法
1. 梅菜洗淨，用清水浸 10 分鐘，擠乾水分，切粒。
2. 梅菜用白鑊炒香，下油 1 湯匙及蒜茸爆香，放入調味料炒勻，盛起備用。
3. 光雞洗淨，抹乾水分，用醃料塗勻雞身內外，將梅菜及薑片釀入雞腔。
4. 熱鍋下油，爆香葱段，下雞隻煎香兩面，傾入清水焗煮約 30 分鐘，至雞隻全熟，煮雞汁留用。
5. 梅菜取出，鋪於碟內；雞稍涼後，斬件，放於梅菜上。
6. 煮雞餘汁用生粉水埋獻，澆在雞件上即成。

◎ Ingredients (Serves 4)
1 chicken (about 1.2 kg)
113 g preserved flowering cabbage
1 tbsp finely chopped garlic
2 slices ginger
2 sprigs spring onion
1.5 cups water

◎ Marinade
1/2 tbsp salt
1/2 tbsp dark soy sauce
1/2 tbsp Shaoxing wine

◎ Seasoning
1/4 tsp salt
1/2 tbsp sugar

◎ Caltrop starch solution
1 tbsp water
1 tsp caltrop starch

Method

1. Rinse the preserved flowering cabbage. Soak in water for 10 minutes. Squeeze water out and dice.

2. Stir-fry the preserved flowering cabbage without oil until fragrant. Add 1 tbsp of oil and stir-fry with garlic until scented. Put in the seasoning and stir-fry. Set aside.

3. Rinse the chicken. Wipe dry. Spread the marinade on the outside and inside of the chicken evenly. Stuff with the preserved flowering cabbage and ginger.

4. Add oil in a heated pot. Stir-fry the spring onion until aromatic. Put in the chicken and fry both sides until fragrant. Pour in the water and cover the lid. Cook for about 30 minutes until it is fully done. Reserve the sauce for later use.

5. Remove the preserved flowering cabbage. Lay on a plate. When the chicken cools down, chop into pieces and place on top of the preserved flowering cabbage.

6. Mix the sauce from cooking the chicken in step (4) with the caltrop starch solution. Sprinkle over the chicken. Serve.

零失敗技巧
Successful Cooking Skills

難以清除梅菜的砂粒，怎辦？
梅菜容易藏有細小的砂粒，應撕開每片梅菜，徹底浸泡及清洗，去掉砂粒才下鍋烹煮。

How to remove the sand grains from the preserved flowering cabbage?
Tear off the leaves and soak them in water. Wash each leaf thoroughly to remove the sand grains.

為何梅菜用白鑊烘炒？
徹底去掉梅菜的水分，而且令梅菜更香口好吃，必須炒至梅菜散發香味啊！

Why stir-fry the preserved flowering cabbage without oil?
This is to dry and make it more fragrant and delicious. Stir-fry until the aroma spreads!

韓式豬軟骨
Korean Style Pork Rib Tips

◎ 材料 （4 至 6 人份量）

豬軟骨 600 克
紅蘿蔔、白蘿蔔各 1 個（小，切件）
洋葱 1 個（切件）
冬菇 5 朵
栗子 8 粒
紅棗 5 粒（去核）
白果數粒
薑茸及乾葱茸各 1 茶匙

◎ 調味料

蜜糖 1 湯匙
清酒 1 湯匙
味醂 1 湯匙
黑椒碎少許

◎ 做法

1. 栗子去殼、去衣；白果去殼、去芯；
   冬菇用水浸軟，去蒂。

2. 豬軟骨飛水，盛起。

3. 燒熱油 1 湯匙，下洋葱、薑茸及
   乾葱茸爆香，加入豬軟骨拌炒，傾
   入水 2 杯煮約 30 分鐘。

4. 下其他配料及調味料，用慢火燜煮
   30 分鐘，至豬軟骨酥脆即可。

◯◯ Ingredients (Serves 4-6)

600 g pork rib tips
1 small carrot (cut into pieces)
1 small white radish (cut into pieces)
1 onion (cut into pieces)
5 dried black mushrooms
8 chestnuts
5 red dates (stoned)
several ginkgos
1 tsp finely chopped ginger
1 tsp finely chopped shallot

◯◯ Seasoning

1 tbsp honey
1 tbsp sake
1 tbsp mirin
crushed black pepper

◯◯ Method

1. Remove the shell and skin of the chestnuts. Remove the shell and core of the ginkgos. Soak the dried black mushrooms in water until soft. Remove the stalks.

2. Scald the pork rib tips. Set aside.

3. Heat up 1 tbsp of oil. Stir-fry the onion, ginger and shallot until fragrant. Add the pork rib tips and stir-fry. Pour in 2 cups of water and cook for about 30 minutes.

4. Put in the other ingredients and seasoning. Simmer for about 30 minutes until the pork rib tips is tender. Serve.

◯◯ 零失敗技巧 ◯◯
Successful Cooking Skills

如何令豬軟骨更好味？
以薑茸、乾葱茸及洋葱起鑊，放入豬軟骨爆香，惹味程度直線上升！

How can I make the pork rib tips even tastier?

Stir-fry finely chopped ginger, shallots and onion in a wok until fragrant. Then sear the pork rib tips over high heat. That would caramelize the meat and make it tasty.

如何有效去掉栗子衣？
我介紹你兩種方法：一. 栗子用水焓 5 分鐘，用布包着栗子，趁熱擦掉外皮。
二. 栗子放入微波爐，加熱 1 至 2 分鐘，令外衣變乾，輕易脫掉。

How to remove the skin of chestnuts effectively?

I suggest two ways: 1. Scald the chestnuts for 5 minutes. Wrap them in a cloth and rub off the skin while warm. 2. Put the chestnuts in a microwave oven and heat up for 1 to 2 minutes. The skin will become dry and easily be stripped off.

牛油蒜茸黑椒煎蝦碌
Fried Prawns with Butter, Garlic and Black Pepper

◎ 材料（3 至 4 人份量）
新鮮大中蝦半斤
蒜茸 1 湯匙
牛油 2 茶匙
紹酒 1 湯匙
粟粉 2 茶匙

◎ 調味料
幼海鹽半茶匙
黑胡椒碎 1 茶匙

◎ 做法

1. 中蝦剪去蝦鬚及蝦腳，挑腸，洗淨，每隻鮮蝦切成兩段，抹乾水分，加入粟粉拌勻。

2. 燒熱鑊下油 2 湯匙，放入中蝦煎至轉成紅色，灒酒炒勻，下蒜茸及調味料炒香，最後加入牛油炒至中蝦汁液收乾即成。

◎ Ingredients (Serves 3-4)

300 g fresh prawns
1 tbsp finely chopped garlic
2 tsp butter
1 tbsp Shaoxing wine
2 tsp cornflour

◎ Seasoning

1/2 tsp fine sea salt
1 tsp crushed black pepper

◎ Method

1. Cut away the tentacles and legs of the prawns with a pair of scissors. Devein and rinse. Cut each prawn into halves. Wipe them dry. Add the cornflour and mix well.

2. Heat up a wok. Add 2 tbsp of oil. Fry the prawns until they turn red. Sprinkle with the Shaoxing wine and stir-fry evenly. Add the garlic and seasoning. Stir-fry until fragrant. Add the butter and stir-fry until the sauce dries. Serve.

◎ 零失敗技巧 ◎
Successful Cooking Skills

為何最後加入牛油拌炒？
令蝦肉帶濃濃的牛油香氣，享用時牛油味濃厚，惹味好吃！
Why add butter in the final step?
It gives the prawns a strong butter flavour, which is sensational!

如何保持蝦肉嫩滑？
毋須剝掉蝦殼，免蝦肉直接受熱，快炒可保持嫩滑質感。
How can I keep the prawns succulent and juicy?
Do not shell them. The shell prevents the flesh from being heated directly. Then fry them quickly to keep them juicy.

 玫瑰香燻雞

Smoked Chicken with Rose and Tea

◎ 材料 （4 至 6 人份量）
冰鮮雞 1 隻（900 克）
乾蔥頭 3 粒（略拍）
薑 6 片（略拍）
蒸雞汁半杯（隔去雞油）

◎ 醃料
鹽 3/4 茶匙
糖半茶匙
紹酒 1 湯匙
生抽及老抽各 1.5 茶匙

◎ 茶燻料
茶葉 1 湯匙
片糖碎 3 湯匙
米 2 湯匙（白鑊炒香）
玫瑰花蕾 2 湯匙

◎ 粟粉水
粟粉 1 茶匙
水 1 湯匙

◎ 做法
1. 冰鮮雞洗淨，抹乾，將醃料擦勻雞身內外。
2. 將乾蔥碎、薑及玫瑰花放入雞腔內，醃 6 小時（或一晚）。
3. 燒滾水，原隻雞隔水蒸 15 分鐘，蒸雞汁留用。
4. 鑊底鋪上錫紙，放上茶燻料，架上蒸架，排上雞，加蓋，用中火燻 15 分鐘，待涼，斬件。
5. 煮滾蒸雞汁，加入適量粟粉水埋獻，澆於雞件上或作為蘸汁伴吃。

◎ Ingredients (Serves 4-6)
1 chilled chicken (900 g)
3 shallots (slightly crushed)
6 slices ginger (slightly crushed)
1/2 cup sauce from the steamed chicken (sieve out chicken oil)

◎ Marinade
3/4 tsp salt
1/2 tsp sugar
1 tbsp Shaoxing wine
1.5 tsp light soy sauce
1.5 tsp dark soy sauce

◎ Ingredients for tea smoking
1 tbsp tea leaves
3 tbsp crushed slab sugar
2 tbsp rice (stir-fried without oil)
2 tbsp rose buds

◎ Cornflour solution
1 tsp cornflour
1 tbsp water

◯◯ Method

1. Rinse the chicken. Wipe dry. Rub the inside and outside of the chicken with the marinade.

2. Put the shallots, ginger and rose buds into the chicken cavity. Marinate for 6 hours (or overnight).

3. Bring water to the boil. Steam the whole chicken for 15 minutes. Reserve the sauce for later use.

4. Lay a wok with aluminum foil. Place the ingredients for tea smoking on the foil. Put a rack on top. Place the chicken on the rack. Put a lid on. Smoke over medium heat for 15 minutes. Leave to cool down. Chop into pieces.

5. Bring the sauce from the steamed chicken to the boil. Thicken with some cornflour solution. Sprinkle on top of the chicken, or serve as a dipping sauce.

◯◯ 零失敗技巧 ◯◯
Successful Cooking Skills

怎樣令燻雞更入味好吃？

緊記將醃料抹勻雞身及雞腔，時間充足，味道自然芳香濃郁！

How can I make the chicken more flavourful?

Make sure you rub the marinade all over the insides and the outsides of the chicken. Allow enough time for the flavours to infuse. The chicken will then be flavourful and aromatic.

用哪款茶葉燻雞？

任何茶葉皆可，我建議用烏龍茶、鐵觀音或普洱茶，香氣濃郁。

Use what kind of tea to make the smoked chicken?

Any kind will do. I suggest using Oolong, Tie Guan Yin, or Puer, all having an intense fragrance.

茶燻的步驟容易控制嗎？

容易辦到！緊記別用大火，以免損壞廚具，而且開動抽油煙機，別弄得廚房煙霧瀰漫！

Is it easy to control the tea smoking steps?

It is easy! Remember not to use high heat to avoid damaging the cookware. Turn on the range hood to prevent heavy smoke from lingering the kitchen!

香茅啤酒雞中翼
Chicken Mid-joint Wings
with Lemongrass in Beer

材料 （4 至 5 人份量）

雞中翼 1 斤
啤酒 1 杯（250 毫升）
香茅 2 棵（切斜片）
蒜頭 2 粒（切片）
指天椒 2 隻（切片）
黑胡椒碎半湯匙
青檸 1 個（榨汁）

醃料

鹽及雞粉各 3/4 茶匙

調味料

魚露 3/4 湯匙
青檸汁 2 茶匙
糖 2 茶匙

做法

1. 雞中翼解凍，洗淨，抹乾水分，加入醃料拌勻醃 20 分鐘。

2. 熱鑊下少許油，放入雞翼煎至兩面微黃色，盛起。

3. 燒熱油，煸炒香茅、蒜頭、指天椒及黑胡椒碎，放入雞翼、啤酒及調味料煮滾，煮約 8 分鐘至雞翼熟透，潷入少許啤酒煮至汁液濃稠，上碟即可。

Ingredients (Serves 4-5)

600 g chicken mid-joint wings
1 cup beer (250 ml)
2 stalks lemongrass (sliced diagonally)
2 cloves garlic (sliced)
2 bird's eye chillies (sliced)
1/2 tbsp crushed black pepper
1 lime (squeezed)

Marinade

3/4 tsp salt
3/4 tsp chicken bouillon powder

Seasoning

3/4 tbsp fish sauce
2 tsp lime juice
2 tsp sugar

Method

1. Defrost the chicken wings. Rinse and wipe dry. Mix with the marinade and rest for 20 minutes.

2. Add a little oil in a heated wok. Fry the chicken wings until both sides turn golden. Set aside.

3. Heat up oil. Sauté the lemongrass, garlic, bird's eye chilli and black pepper. Put in the chicken wings, beer and seasoning. Bring to the boil. Cook for about 8 minutes until the chicken wings are cooked through. Sprinkle with a little beer. Cook until the sauce thickens. Serve.

◎ 零失敗技巧 ◎
Successful Cooking Skills

1 斤雞中翼約有多少隻？

體型小的雞翼約 15 隻；較大的數量會略少。

How many chicken mid-joint wings we can get for 600 g?

We can have about 15 pieces for small ones but fewer for bigger ones.

何謂煸炒？

先將鑊大火加熱，放入少量油，下食材慢火受熱，迫出香氣！

What does "sauté" mean?

This is a stir-frying process in which a wok is heat up over high heat, and then a little oil is added to stir-fry the food ingredients over low heat. The ingredients will be heated up and release their fragrance.

如何增添啤酒的香氣？

啤酒遇熱後容易揮發，故上碟前再潷入少許啤酒，令啤酒香氣更持久。

How to enhance the aromatic flavour of beer?

Heated beer is easy to evaporate. To keep the fragrance long, sprinkle with a little beer before serving.

愛香茅的清香，如何有效散發香味？

先用刀略拍香茅，再切成斜片，有效地散發香茅的清香味道，令餸菜惹味好吃！

The light fragrance of lemongrass is lovely. How to make it fully spread?

Slightly crush the lemongrass with a knife and then slice diagonally. It helps the lemongrass to release its fresh aroma, making the dish palatable!

香草焗豬手

Baked Pork Knuckle with Herbs

◎ 材料 （4 人份量）
包裝鹹豬手 1 隻（未調味）
薑 3 片
黑椒碎 2 茶匙
混合乾香草適量

◎ 做法

1. 鹹豬手略洗，備用。

2. 燒滾水，放入薑片及鹹豬手煮滾，轉中小火煲 1 小時，盛起豬手待片刻。用黑椒碎抹勻豬手，灑入適量混合乾香草。

3. 將已醃味的豬手放在焗盤，放入已預熱之焗爐，用 180℃ 上下火焗約 40 分鐘，待豬手表皮呈金黃色及散發陣陣烤肉香味，取出，待 15 分鐘，切塊享用。

Ingredients (Serves 4)
1 packed salt-cured pork knuckle (not spiced)
3 slices ginger
2 tsp crushed black pepper
mixed dried herbs

Method
1. Slightly rinse the pork knuckle. Set aside.
2. Bring water to the boil. Put in the ginger and pork knuckle. Bring to the boil. Turn to low-medium heat and cook for 1 hour. Dish up. Leave for a while. Rub the pork knuckle with the crushed black pepper. Sprinkle with some mixed dried herbs.
3. Put the spiced pork knuckle into a baking tray. Bake in a preheated oven with both upper and lower heat at 180°C for about 40 minutes. When the pork knuckle gives a golden outside with a roast meat fragrance, take out and leave for 15 minutes. Cut into pieces and serve.

零失敗技巧
Successful Cooking Skills

想吃濃濃的肉香味，如何辦？
可減省黑椒碎及混合香草的份量，以免過度掩蓋豬手之肉香。
How to make the meat flavour strong?
Reduce the quantity of crushed black pepper and mixed herbs to avoid overpowering.

家裏只有小烤爐，怎辦？
若放不下整隻豬手，可斬件烤焗。由於發熱線貼近，建議蓋上錫紙以免焦燶。
There is only a small oven at home. How to do?
If there is not enough room for the whole pork knuckle, chop it into pieces before baking. As the heating wire is near, it is better to cover the pork knuckle with aluminum foil to avoid scorching.

京葱孜然羊肉片

Lamb and Peking Scallion with Cumin Powder

◎ 材料 （4 人份量）
急凍羊肩肉或羊髀肉 300 克
京葱 1 棵（切斜段）
蒜茸 2 茶匙
孜然粉及紅椒粉各 1/3 茶匙
荷葉餅 6 塊

◎ 醃料
鹽半茶匙
生抽 2 茶匙
糖半茶匙
胡椒粉及麻油各少許
紹酒 1 茶匙
蛋白 1 湯匙
孜然粉 3/4 茶匙
生粉 2 茶匙（後下）

◎ 做法

1. 羊肩肉放於雪櫃下層自然解凍，洗淨，順橫紋切片，用醃料拌勻。

2. 燒熱少許油，下京葱炒至軟身，灑入鹽 1/4 茶匙炒勻，盛起。

3. 燒熱油 1 湯匙，下蒜茸及羊肉片，略煎兩面後推散，用中大火炒至乾身，盛起。

4. 最後灑上孜然粉及紅椒粉，用荷葉餅夾羊肉及京葱伴吃。

(◎) Ingredients (Serves 4)

300 g frozen lamb shoulder chops or lamb leg

1 stalk Peking scallion (cut diagonally into sections)

2 tsp finely chopped garlic

1/3 tsp cumin powder

1/3 tsp paprika

6 lotus leaf-shaped pancakes

(◎) Marinade

1/2 tsp salt

2 tsp light soy sauce

1/2 tsp sugar

ground white pepper

sesame oil

1 tsp Shaoxing wine

1 tbsp egg white

3/4 tsp cumin powder

2 tsp caltrop starch (added at last)

(◎) Method

1. Defrost the lamb shoulder chops in the lower chamber of the refrigerator. Rinse and slice. Cut across the grains of lamb. Mix the lamb with the marinade.

2. Heat up a little oil. Stir-fry the Peking scallion until soft. Sprinkle with 1/4 tsp of salt and stir-fry. Remove.

3. Heat up 1 tbsp of oil. Slightly fry the garlic and lamb on both sides. Scatter and stir-fry over medium-high heat until dry. Set aside.

4. Sprinkle with cumin powder and paprika. Wrap in the pancakes with the Peking scallion to serve.

(◎) 零失敗技巧 (◎)
Successful Cooking Skills

此菜的烹調重點如何？

菜式完成時，醬汁黏着羊肉，別炒至水汪汪，醬汁乾身是此菜之重點。

Is there anything that needs my attention in this recipe?

The key is the consistency of the sauce, which should cling on to the lamb. It should not be watery.

最後羊肉用中大火炒至乾身，為甚麼？

醬汁慢慢收乾，濃縮調味的精華，令味道更濃厚！

The lamb is stir-fried over medium-high heat until dry in the final step. Why?

This is to let the seasoning condense by reducing the sauce slowly, giving the lamb a more intense flavour!

 一蝦兩吃

Deep-fried Prawns with Wasabi Sauce and Kumquat Syrup

◎ **材料 （4 人份量）**
大蝦 8 隻

◎ **醃料**
鹽 1/8 茶匙
胡椒粉少許

◎ **炸漿料**
蛋黃 2 個
水 1 湯匙
炸粉 2 湯匙

◎ **日式芥辣汁**
日式芥辣 1 湯匙
蜜糖 4 茶匙
水 4 湯匙
鹽 1/8 茶匙

◎ **柑桔蜜汁**
蜜餞柑桔 3 粒（去核、剁碎）
檸檬汁半湯匙
糖 2 茶匙
水 3 湯匙

◎ **做法**

1. 大蝦去殼，開背、去腸，切成蝦球，洗淨及抹乾，加入醃料醃 15 分鐘。

2. 蛋黃拂勻，拌入其餘炸漿料，備用。

3. 大蝦蘸上炸漿，放入熱油炸至金黃色及熟透，分成兩份，排於碟上。

4. 分別煮滾日式芥辣汁及柑桔蜜汁，伴大蝦享用。

Ingredients (Serves 4)
8 prawns

Marinade
1/8 tsp salt
ground white pepper

Deep-frying batter
2 egg yolks
1 tbsp water
2 tbsp deep-frying powder

Washabi sauce
1 tbsp washabi
4 tsp honey
4 tbsp water
1/8 tsp salt

Kumquat syrup
3 sweetened kumquat (cored and chopped)
1/2 tbsp lemon juice
2 tsp sugar
3 tbsp water

Method
1. Shell the prawns. Cut open at the back and devein. Rinse and wipe dry. Marinate for 15 minutes.

2. Whisk egg yolks and mix in the remaining ingredients of deep-frying batter. Set aside.

3. Coat prawns with the deep-frying batter. Deep-fry in hot oil until golden brown and done. Divide into 2 portions and arrange on a plate.

4. Heat the washabi sauce and kumquat syrup respectively. Serve with the prawns.

一蝦兩吃

零失敗技巧
Successful Cooking Skills

如何炮製鮮甜爽口的蝦？
建議選購新鮮活蝦，鮮甜爽口，美味無窮！
How to cook sweet and crunchy prawns?
Choose fresh and live prawns that taste sweet and crunchy.

甚麼是蜜餞柑桔？
涼果店有售之蜜餞柑桔，味道甜膩，製成汁很醒胃。
What is sweetened kumquat?
It is sold at dried fruits stores and tastes super sweet.

香醋子薑雞

Braised Chicken with Pickled Young Ginger in Sweetened Black Vinegar

◎ 材料 （4 人份量）
光雞半隻（12 兩）
黑甜醋 1 杯（250 毫升）
酸子薑 4 兩（見 p.57）

◎ 做法
1. 光雞洗淨，斬件，放入白鑊炒乾水分。
2. 煮滾黑甜醋，放入雞件待滾，轉慢火煮約10分鐘至雞肉熟透，關火，浸1小時，最後加入酸子薑拌勻，上碟品嘗。

◎ Ingredients (Serves 4)
1/2 chicken (450 g)
1 cup sweetened black vinegar (250 ml)
150 g pickled young ginger (see p.57)

◎ Method
1. Rinse the chicken. Chop into pieces. Stir-fry without oil until dry.
2. Bring the sweetened black vinegar to the boil. Put in the chicken and bring to the boil. Turn to low heat and simmer for about 10 minutes until fully cooked. Turn off heat. Soak for 1 hour. Finally add the pickled young ginger and mix well. Serve.

◎ 零失敗技巧 ◎
Successful Cooking Skills

用新鮮雞及冰鮮雞炮製，食味有很大分別嗎？
新鮮雞鮮嫩滑溜，雞味濃重；若想便宜一點，冰鮮雞也是不錯之選擇。
Is there any difference in flavour between fresh chicken and chilled chicken?
Smooth and tender, fresh chicken has a rich chicken flavour. Inexpensive chilled chicken is also a good choice.

雞件為何用白鑊先炒乾？
目的是去掉雞肉的水分，再注入黑甜醋浸煮，保留黑甜醋之原汁原味！
Why stir-fry the chicken without oil beforehand?
This is to take the moisture out of the chicken and to keep the original flavour of black vinegar cooked with the chicken.

甚麼是黑甜醋？
即坐月子用的醋，香甜味濃，醋香引人，而且具補身功效。
What is sweetened black vinegar?
That is the vinegar used by women after child delivery for nourishing. It has a rich sweet flavour and benefits the body.

* 酸子薑簡易醃法
*Easy method for pickled young ginger

◎ 材料
子薑 8 兩
鹽 1 湯匙

◎ 糖醋汁
白醋 10 湯匙
汕頭米醋 2 湯匙
糖 12 湯匙

◎ 做法
1. 子薑用小刀刮去外皮，洗淨，切片（厚薄均勻）。
2. 子薑用鹽 1 湯匙醃 25 分鐘，沖淨，吸乾水分，吹至乾爽。
3. 糖醋汁煲至糖完全溶化，待涼，放入子薑浸 2 小時即可。

◎ Ingredients
300 g young ginger
1 tbsp salt

◎ Sugar and vinegar solution
10 tbsp white vinegar
2 tbsp Shantou rice vinegar
12 tbsp sugar

◎ Method
1. Scrape the young ginger with a small knife. Rinse and slice (in even thickness).
2. Marinate the ginger with 1 tbsp of salt for 25 minutes. Rinse, wipe dry and air-dry.
3. Cook the sugar and vinegar solution until the sugar melts. Allow it to cool down. Put in the young ginger and soak for 2 hours. Ready to serve.

梅膏三文魚

Salmon in Plum Paste

◎ 材料 （4 人份量）
急凍或冰鮮三文魚 350 克
梅膏 2 湯匙
乾葱茸及紅椒粒各 2 湯匙
蛋汁 1 湯匙
生粉 2 湯匙

◎ 醃料
鹽半茶匙
黑椒粉少許

◎ 做法
1. 急凍三文魚放於雪櫃下層自然解凍，洗淨，抹乾水分。
2. 三文魚去皮、去骨，切件，下醃料拌勻。
3. 三文魚塊與蛋汁拌勻，沾上生粉，放入熱油半煎炸至金黃色，盛起待一會，再半煎炸片刻，吸取多餘油分。
4. 燒熱少許油，加入乾葱茸及紅椒粒爆香，下梅膏及調味料煮片刻，拌入三文魚塊，裹上醬汁即可上碟。

⬭ Ingredients (Serves 4)

350 g frozen or chilled salmon
2 tbsp plum paste
2 tbsp finely chopped shallot
2 tbsp diced red chilli
1 tbsp egg wash
2 tbsp caltrop starch

⬭ Marinade

1/2 tsp salt
ground black pepper

⬭ Method

1. Defrost the salmon in the lower chamber of the refrigerator. Rinse and wipe dry.

2. Skin and bone the salmon. Cut into pieces. Mix with the marinade.

3. Mix the salmon with the egg wash. Coat with the caltrop starch. Fry and deep-fry in hot oil until golden. Leave for a while. Fry and deep-fry again for a moment. Drain the oil.

4. Heat up a little oil. Stir-fry the shallot and red chilli until scented. Add the plum paste and seasoning. Cook for a while. Mix in the salmon. When it is coated with the sauce, dish up and serve.

⬭ 零失敗技巧 ⬭
Successful Cooking Skills

甚麼是梅膏？
梅膏是用酸梅等煮成，酸甜美味，在潮州雜貨店有售。

What is plum paste?
Plum paste is made of pickled plums and other condiments. It is delicious with a sweet and sour taste, and is available at Chaozhou food groceries.

可自製梅膏嗎？
絕對可以！準備酸梅 12 粒、冰糖 100 克及水半杯。先將酸梅去核、弄碎；冰糖舂碎；將全部材料煮至濃稠，隔渣，可儲存使用。

Can I make plum paste myself?
Absolutely! Prepare 12 pickled plums, 100 g rock sugar and 1/2 cup of water. Core and mash the pickled plums. Crush the rock sugar. Cook all the ingredients until the sauce is thicken. Sieve the sauce. It can be stored for future use.

羅望子醬馬鈴薯燜鴨

Braised Duck with Potato in Tamarind Paste

材料 （4 人份量）

冰鮮鴨半隻
馬鈴薯 2 個
羅望子半塊
薑 4 片
蒜肉 6 粒
麵豉醬 2 湯匙
芫茜 4 棵

醃料

生抽 2 茶匙
紹酒 2 茶匙

調味料

老抽 2 茶匙
片糖半片

做法

1. 羅望子用滾水 1 量杯浸軟，挑去果核備用。

2. 馬鈴薯去皮，洗淨，切塊，用清水浸過面。

3. 冰鮮鴨洗淨，抹乾水分，將醃料抹勻鴨皮及內腔。

4. 將鴨放入油鑊煎至表皮金黃色，盛起。原鑊留下油 3 湯匙，下薑片、蒜肉及馬鈴薯炒香。

5. 放入鴨、羅望子醬、調味料及麵豉醬，傾入滾水（宜浸過鴨面）及芫茜煮滾片刻，轉慢火燜約 1 小時至鴨肉軟腍，斬件食用。

Ingredients (Serves 4)

1/2 frozen duck
2 potatoes
1/2 slab dried tamarind
4 slices ginger
6 cloves skinned garlic
2 tbsp fermented soybean paste
4 sprigs coriander

Marinade

2 tsp light soy sauce
2 tsp Shaoxing wine

Seasoning

2 tsp dark soy sauce
1/2 raw cane sugar slab

Method

1. Soak dried tamarind in 1 cup of hot water until soft. Remove the seeds. Set aside.

2. Peel, rinse and cut potatoes into chunks. Soak them in water completely.

3. Rinse the duck and wipe dry. Brush the marinade on the skin and the inside of the duck.

4. Heat oil in a wok. Fry the whole duck in oil until golden. Set aside. Keep 3 tbsp of oil in the same wok. Stir fry ginger, garlic and potatoes until fragrant.

5. Put in the duck and add tamarind paste from step (1). Pour in seasoning and fermented soybean paste. Add water to cover the duck. Sprinkle coriander on top and boil for a while. Turn to low heat and simmer for about 1 hour until the duck is tender. Chop it up and serve.

◎ 零失敗技巧 ◎
Successful Cooking Skills

怎樣處理羅望子？
羅望子用水浸軟後，必須隔渣，去核及去皮，令醬汁濃滑及美味。
How to prepare dried tamarind?
Make sure you strain the paste after soaking with water. The seeds and skin in dried tamarind would make your sauce gritty.

馬鈴薯為何用清水浸過面？
以免馬鈴薯接觸空氣後氧化，令色澤變黑，而且浸水後的馬鈴薯，肉質更軟糯。
Why do you add water to cover the potatoes?
Potatoes will turn black once get in touch with air. Soaking them in water prevents it from darkening. Also, the potatoes tend to be mushier and creamier after being soaked in water.

加入麵豉醬燜煮，有何作用？
加入麵豉醬燜鴨肉，可提升肉香鮮味，鴨香惹味！
Why do you add fermented soybean paste when you stew the duck?
It brings out the meaty flavour of the duck and makes it taste better.

為何用片糖作調味料？
燜煮菜式宜用片糖調味，甜味比砂糖更豐富，長時間燜煮，食味更佳！
Why do you season this dish with raw cane sugar slabs?
Braised or stewed dishes usually call for raw cane sugar slabs because its sweetness has more depth and richness than white sugar. It tastes even better after prolonged cooking.

自製甜椒醬伴鮫魚
Fried Mackerel with Bell Pepper Sauce

◎ 材料 （4 人份量）
鮫魚 4 件（約 8 兩）
甜青椒 4 兩
甜紅椒 4 兩
紫洋葱 1.5 兩
清水 150 毫升

◎ 醃料
鹽半茶匙
胡椒粉少許
生粉 1 湯匙（後下）

◎ 調味料
茄汁 1.5 湯匙
是拉差辣椒醬半湯匙
鹽 1/3 茶匙
糖 1 湯匙

◎ 做法
1. 鮫魚洗淨，抹乾水分，與醃料拌勻醃 10 分鐘。
2. 青紅甜椒及紫洋葱各 1/4 份量切絲，其餘切粒。
3. 燒熱適量油，下青紅椒絲及紫洋葱絲炒透，盛起。
4. 下青紅椒粒及洋葱粒炒透，放入攪拌機，傾入 1/3 份量清水打成甜椒茸，備用。
5. 魚塊均勻地灑上生粉，放入熱油內煎熟，上碟。
6. 煮滾甜椒茸、調味料及餘下之清水，加入青紅椒絲及洋葱絲拌勻，澆在魚塊上享用。

◎ Ingredients (Serves 4)
4 pieces mackerel (about 300 g)
150 g green bell pepper
150 g red bell pepper
57 g purple onion
150 ml water

◎ Marinade
1/2 tsp salt
ground white pepper
1 tbsp caltrop starch (added at last)

◎ Seasoning
1.5 tbsp ketchup
1/2 tbsp Sriracha chilli sauce
1/3 tsp salt
1 tbsp sugar

◎ Method
1. Rinse mackerel and wipe dry. Marinate for 10 minutes.
2. Shred 1/4 portion of green bell pepper, red bell pepper and purple onion. Dice the remaining portions.
3. Heat oil in a wok. Add shredded green and red bell peppers and purple onion shreds. Stir-fry well and set aside.
4. Stir-fry the diced green and red bell peppers and diced purple onion thoroughly. Put into a blender and add 1/3 portion of water. Blend into minced pepper.
5. Coat fish evenly with caltrop starch. Fry in hot oil until done. Put into a plate.
6. Bring the minced pepper, seasoning and the remaining water to the boil. Mix in the green and red bell peppers shreds as well as the purple onion shreds. Pour the sauce over the fish and serve.

◯◯ 零失敗技巧 ◯◯
Successful Cooking Skills

除鮫魚外，還可選擇哪種魚？

銀鱈魚及三文魚也是不錯之選擇，而且營養價值非常高。

What fish can be used except mackerel?

White cod fish and salmon are good choices as they are high in nutritional value.

步驟似乎很繁複，容易掌握嗎？

其實步驟並不複雜，只要將青紅甜椒及洋蔥分成 3/4 份及 1/4 份，前者製成椒茸；後者作為配料即可。

Would it be easy to handle these rather complicated steps?

In fact the steps are not complex. Just divide green and red bell peppers as well as the purple onion into 3/4 portion and 1/4 portion; and the former is used to make minced pepper while the latter is used as condiments.

甚麼是「是拉差辣椒醬」？

這是一款泰式辣椒醬，辣味濃重，於泰式雜貨店有售。

What is Sriracha chilli sauce?

This is a kind of Thai chilli sauce which has strong spicy flavor and can be bought from Thai grocery stores.

橙酒焗春雞

Roasted Spring Chicken with Orange Wine

材料（3 至 4 人份量）
春雞 1 隻（約 800 克）
甜鮮橙 1 個（大）
橙酒 4 湯匙

醃料
鹽 3/4 茶匙
生抽 1 湯匙
糖 1 茶匙
橙汁 2 湯匙

調味料
水 1 杯
鹽半茶匙
糖 4 茶匙
老抽半湯匙

做法
1. 鮮橙刮出橙皮絲半湯匙；輕榨橙汁 2 湯匙，備用。
2. 春雞洗淨，拔淨幼毛，抹乾雞身及雞腔。
3. 醃料拌勻，塗抹在春雞內外醃 1.5 小時。
4. 熱鍋下油，放入春雞煎至微黃色，傾入調味料焗煮約 20 分鐘，灒橙酒，再煮 5 分鐘至汁液濃稠，待雞隻稍涼，斬件上碟。
5. 最後澆上餘下的汁料，以橙皮絲裝飾即可。

Ingredients (Serves 3-4)
1 spring chicken (about 800 g)
1 large sweet orange
4 tbsp orange wine

Marinade
3/4 tsp salt
1 tbsp light soy sauce
1 tsp sugar
2 tbsp orange juice

Seasoning
1 cup water
1/2 tsp salt
4 tsp sugar
1/2 tbsp dark soy sauce

Method
1. Make 1/2 tbsp of shredded orange zest. Squeeze 2 tbsp of juice from the orange. Set aside.
2. Rinse and pluck tiny hairs from the chicken. Wipe it dry both outside and inside.
3. Mix the marinade. Spread on the outside and inside of the chicken. Rest for 1.5 hours.
4. Add oil in a heated pot. Fry the chicken until light brown. Pour in the seasoning and cook with a lid on for 20 minutes. Sprinkle with the orange wine. Cook for another 5 minutes until the sauce reduces. Chop up the chicken when it cools down. Place on a plate.
5. Sprinkle the remaining sauce on top. Decorate with the orange zest and serve.

◎ 零失敗技巧 ◎
Successful Cooking Skills

用鍋焗煮春雞，比用焗爐有何分別？

用鍋烹調的春雞，肉質較濕潤，雞肉不會太乾；用焗爐烤烘的春雞，外皮香脆！

What is the difference between cooking the chicken in a pot and in an oven?

The chicken will have a moister texture by cooking in a pot whereas baking the chicken in an oven will bring a crunchy outside.

哪裏購買橙酒？

大型超級市場、酒行及烘焙材料供應店皆可找到；若不想浪費，建議購買小瓶裝的酒辦！

Where to buy orange wine?

It is available at supermarkets, wine shops, and shops for baking ingredients. If you don't want to waste any wine, buy miniature liquor instead.

春雞是貴價食材嗎？

不是！一般急凍食品店皆有出售；若想試試品質較高的春雞，可到日式超市購買，但價錢當然略貴。

Is the spring chicken expensive?

No. It can be bought at frozen food shops. If you want a quality spring chicken, buy one in the Japanese supermarket, which is more expensive.

糖醋魚塊

Fish Fillets in Sweet Vinegar Sauce

◎◎ 材料 （4 人份量）
急凍龍脷柳 2 塊
甜青椒及番茄粒 1 湯匙
雞蛋半個（拂勻）
生粉 1/4 杯
松子仁 1 湯匙

◎◎ 醃料
鹽 1/3 茶匙
胡椒粉少許

◎◎ 糖醋汁（拌勻）
鎮江醋 4 湯匙
片糖碎 3 湯匙
茄汁 3 湯匙
鹽 1/4 茶匙
喼汁 1 湯匙
水 4 湯匙

糖醋魚塊

◎◎ 生粉水（拌勻）
生粉 2 茶匙
水 1 湯匙

◎◎ 做法
1. 龍脷魚柳放於雪櫃下層自然解凍，洗淨，用乾布抹乾，切件，加入醃料拌勻。
2. 燒熱水 3/4 杯，灑入松子仁及糖 2 茶匙浸片刻，盛起，吹乾，放入焗爐烤至香脆（或用少許油炸至金黃色），備用。
3. 龍脷魚塊拌入蛋汁，裹上生粉，放入熱油半煎炸至微黃色，盛起待涼，再半煎炸一次，盛起。
4. 燒熱少許油，加入糖醋汁煮約 5 分鐘，拌入適量生粉水埋獻，下青椒粒及番茄粒，將汁料澆於魚塊上，最後灑上松子仁即可。

◎◎ Ingredients (Serves 4)
2 frozen sole fillets
1 tbsp diced green bell pepper
and diced tomato
1/2 egg (beaten)
1/4 cup caltrop starch
1 tbsp pine nuts

◎◎ Marinade
1/3 tsp salt
ground white pepper

◎◎ Sweet vinegar sauce (mixed well)
4 tbsp Zhenjiang vinegar
3 tbsp crushed slab sugar
3 tbsp ketchup
1/4 tsp salt
1 tbsp Worcestershire sauce
4 tbsp water

◎◎ Caltrop starch solution (mixed well)
2 tsp caltrop starch
1 tbsp water

Method

1. Defrost the sole fillets in the lower chamber of the refrigerator. Rinse and wipe dry with a dry cloth. Cut into pieces. Mix with the marinade.

2. Heat up 3/4 cup of water. Put in the pine nuts and 2 tsp of sugar. Soak for a moment. Dish up and air-dry. Bake in an oven until crunchy (or deep-fry with a little oil until golden). Set aside.

3. Mix the sole fillets with the egg wash. Coat with the caltrop starch. Put in hot oil to fry and deep-fry until light brown. Leave to cool down. Fry and deep-fry again. Set aside.

4. Heat up a little oil. Add the sweet vinegar sauce and cook for about 5 minutes. Thicken the sauce with some caltrop starch solution. Put in the green bell pepper and tomato. Pour the sauce onto the fish fillets. Finally sprinkle the pine nuts on top.

零失敗技巧
Successful Cooking Skills

如何吃得香脆的魚塊？
先將魚塊蘸上蛋白，再撲上厚厚的生粉，炸後的魚塊，香脆可口！

How to make fragrant and crisp fish fillets?

Dip the fish fillets in egg white and then coat with a thick layer of caltrop starch. After deep-frying, you can taste crunchy and delectable fish fillets!

為何將魚肉徹底抹乾？
急凍魚肉解凍後會分泌很多水分，抹乾後才上粉炸，吃起來魚塊更乾、更香脆！

Why wipe the fish thoroughly dry?

Frozen fish will release a lot of water after it is defrosted. Drying it before coating and deep-frying makes it drier and more crispy!

生粉及粟粉有何分別？用粟粉可以嗎？
作為醃料或埋獻之用，生粉或粟粉皆可；但用作上粉油炸的話，以生粉的效果較香脆。

What is the difference between caltrop starch and cornflour? Can we use cornflour?

Caltrop starch or cornflour can be used for marinating food or thickening sauce. As for deep-frying, food coated with caltrop starch is crispier.

金不換辣椒膏炒青口
Stir-fried Mussels with Thai Basil and Chilli Paste

◎ 材料 （6 人份量）

急凍青口 600 克

金不換 2 棵

香茅 1 枝（切碎）

紅辣椒半隻（去籽、切碎）

蒜茸 2 茶匙

泰式辣椒膏 1 湯匙

◎ 調味料

鹽 1/3 茶匙

糖半茶匙

魚露 2 茶匙

胡椒粉少許

水 4 湯匙

◎ 做法

1. 金不換摘出葉片，去莖。

2. 急凍青口放於雪櫃下層自然解凍，放入沸水燙至略開口，盛起。

3. 燒熱油 1 湯匙，加入蒜茸、紅辣椒及香茅炒香，下青口炒勻。

4. 加入辣椒膏及調味料炒勻，加蓋，焗煮片刻。

5. 最後加入金不換葉略炒，上碟享用。

Ingredients (Serves 6)

600 g frozen mussels
2 stalks Thai basil
1 stalk lemongrass (chopped)
1/2 red chilli (deseeded; chopped up)
2 tsp finely chopped garlic
1 tbsp Thai chilli paste

Seasoning

1/3 tsp salt
1/2 tsp sugar
2 tsp fish sauce
ground white pepper
4 tbsp water

Method

1. Pick the leaves of the Thai basil. Discard the stems.

2. Defrost the mussels in the lower chamber of the refrigerator. Blanch until the shells open a little bit. Set aside.

3. Heat up 1 tbsp of oil. Stir-fry the garlic, red chilli and lemongrass until fragrant. Add the mussels and stir-fry evenly.

4. Put in the chilli paste and seasoning. Stir-fry. Put a lid on and cook for a moment.

5. Finally add the Thai basil leaves and slightly stir-fry. Serve.

零失敗技巧
Successful Cooking Skills

只取用金不換葉嗎？
是啊！金不換的葉片散發濃烈的香氣，是炒煮青口的最後拍檔。

Only basil leaves are used?

Yes! Basil leaves bring off an intense aroma. They are the last condiment to be cooked with the mussels.

青口為何先用沸水灼一會？
能挑出未能開口已死掉的青口。

Why blanch mussels for a while first?

It helps pick out dead mussels with closed shells.

 口水雞

Steamed Chicken Dressed in Sichuan Peppercorn Chilli Oil

材料 （4 人份量）

冰鮮雞半隻
即食海蜇 4 兩
炒香白芝麻 2 茶匙
薑 4 片
芫茜適量

醃料

鹽 1 茶匙
紹酒 1 湯匙

醬汁（拌勻）

麻香辣椒油 3 茶匙（見 p.76）
麻油 2 茶匙
鎮江醋 1 湯匙
糖 1 茶匙
鹽 1/3 茶匙

做法

1. 雞洗淨，抹乾水分，下醃料塗勻醃
   1 小時。薑片放於雞上，隔水大火
   蒸 20 分鐘，待涼。

2. 雞斬件上碟，伴即食海蜇，澆上醬
   汁，灑入炒香白芝麻及芫茜供食。

Ingredients (Serves 4)

1/2 chilled chicken
150 g instant jellyfish
2 tsp toasted white sesames
4 slices ginger
coriander

Marinade

1 tsp salt
1 tbsp Shaoxing wine

Dressing (mixed well)

3 tsp Sichuan peppercorn chilli oil (see p.76)
2 tsp sesame oil
1 tbsp Zhenjiang black vinegar
1 tsp sugar
1/3 tsp salt

Method

1. Rinse the chicken and wipe dry. Add marinade and brush well all over the skin. Arrange the sliced ginger over the chicken. Steam over high heat for 20 minutes. Leave it to cool.

2. Chop the chicken into pieces. Arrange instant jellyfish on the side. Dribble the dressing all over. Sprinkle toasted sesames on top. Garnish with coriander. Serve.

* 麻香辣椒油
*Sichuan Peppercorn Chilli Oil

◎ 材料

川椒粒 4 湯匙
指天椒 4 兩
豆豉 1 湯匙
蝦米 2 湯匙
乾葱茸 2 湯匙
蒜茸 1 湯匙
粟米油 1.5 杯

◎ 調味料

鹽 1 茶匙
糖 1 茶匙
生抽 1 湯匙

◎ 做法

1. 蝦米洗淨，切碎；豆豉用水沖洗，切碎；指天椒洗淨，去蒂、切碎。

2. 燒熱油 1.5 杯，下川椒粒用小火炸至散發香味，隔去大部份川椒粒，下蝦米、指天椒及乾葱茸，用小火炒至帶香味，加入豆豉及蒜茸炒勻，最後下調味料煮 5 分鐘（整個炒煮過程約 15 至 20 分鐘），待涼，入瓶儲存。

◎ Ingredients

4 tbsp Sichuan peppercorns
150 g bird's eye chillies
1 tbsp fermented black beans
2 tbsp dried shrimps
2 tbsp finely chopped shallot
1 tbsp finely chopped garlic
1.5 cups corn oil

◎ Seasoning

1 tsp salt
1 tsp sugar
1 tbsp light soy sauce

◎ Method

1. Rinse the dried shrimps and chop them finely. Rinse the fermented black beans with water and chop them finely. Rinse the bird's eye chillies. Cut off their stems and chop finely.

2. Heat a wok and pour in 1.5 cup of oil. Fry the Sichuan peppercorns over low heat until fragrant. Set aside most of the Sichuan peppercorns with a strainer ladle. Add dried shrimps, bird's eye chillies and shallot. Stir fry over low heat until fragrant. Add fermented black beans and finely chopped garlic. Stir well. Add seasoning at last and cook for 5 minutes. (The whole cooking step takes about 15 to 20 minutes.) Leave it to cool. Transfer into sterilized bottle.

◎ 零失敗技巧 ◎
Successful Cooking Skills

炒煮辣椒油有何要點？

炒川椒粒時必須注意火候，用小火炒透，別冒出大煙及炒焦，時間太久令油帶苦澀味。

What is the key for making the chilli oil?

Secret tricks: When you stir fry the Sichuan peppercorns, pay attention to the heat you use. Always fry them over low heat. Don't let them smoke and don't burn them. If you fry them for too long the chilli oil will end up tasting bitter.

怎樣蒸雞才嫩滑好吃？

按食譜的方法蒸 20 分鐘，關火，再焗 10 分鐘，徹底焗透可保持肉質鮮嫩。

How can you steam the chicken to perfection?

Just steam for 20 minutes according to the recipe. Turn off the heat and leave the chicken in the wok or steamer for 10 minutes, with the lid covered. The gentle remaining heat will cook through the chicken without overcooking it. That's the key to tender and succulent chicken meat.

口水雞有何特色？

口水雞味麻辣、帶醋香味，伴飯吃，滋味無窮！

What is so special about this recipe?

It tastes numbing and spicy, with a hint of vinegar fragrance. It goes well with rice.

咖喱三文魚頭粉皮煲

Simmered Salmon Head and Vermicelli Sheet with Curry Sauce in Clay Pot

○ 材料 （4 至 5 人份量）

三文魚頭 1 個（約 1 斤 4 兩）
即食粉皮 1 包
咖喱醬 1.5 湯匙
咖喱粉 1 茶匙
乾葱頭 4 粒
椰漿半杯（125 毫升）
紅辣椒 1 隻（切圈，裝飾用）

○ 醃料

鹽 1 茶匙
胡椒粉少許
薑汁酒 1 湯匙
生粉 1.5 湯匙（後下）

○ 調味料

水 1.5 杯
魚露 2 湯匙
胡椒粉少許
糖 1 茶匙

○ 做法

1. 三文魚頭洗淨，斬件，與醃料拌勻
   醃 15 分鐘。

2. 粉皮隔去水分，備用。

3. 燒熱適量油，魚頭抹上生粉，放入
   油鍋內炸透，盛起。

4. 瓦鍋內燒熱少許油，下乾葱頭、
   咖喱醬及咖喱粉爆香，加入魚頭
   及調味料煮滾，燜煮約 10 分鐘，
   最後下粉皮及椰漿煮滾，裝飾後
   上桌品嘗。

○ Ingredients (Serves 4-5)

1 salmon head (about 750 g)
1 pack instant vermicelli sheet
1.5 tbsp curry paste
1 tsp curry powder
4 shallots
1/2 cup (125 ml) coconut milk
1 red chilli (cut into rings for garnishing)

○ Marinade

1 tsp salt
ground white pepper
1 tbsp ginger wine
1.5 tbsp caltrop starch (added at last)

○ Seasoning

1.5 cups water
2 tbsp fish sauce
ground white pepper
1 tsp sugar

○ Method

1. Rinse salmon head and chop into
   pieces. Marinate for 15 minutes.

2. Drain vermicelli sheet and set aside.

3. Heat oil in a wok. Coat the salmon
   head with caltrop starch and deep-
   fry until done. Drain.

4. Heat a little oil in a clay pot. Stir-fry
   shallots, curry paste and curry powder
   until fragrant. Put in the fish head
   and seasoning. Bring to the boil and
   simmer for about 10 minutes. Add the
   vermicelli sheet and coconut milk.
   Bring to the boil. Garnish and serve.

◎ 零失敗技巧 ◎
Successful Cooking Skills

這道菜辣味濃郁嗎？
此菜屬於中級辣味，卻滲有陣陣濃烈的咖喱香味，令人食指大動。

Is this dish spicy?
This dish belongs to medium spicy but it has strong curry fragrance.

粉皮毋須處理嗎？
粉皮呈軟身，即開即煮，於大型凍肉食品店有售。

Is it necessary to handle the vermicelli sheet?
Vermicelli sheet is soft and is ready for cooking. It can be bought from large frozen food stores.

用瓦鍋烹煮，有何注意之處？
先用慢火令瓦鍋略預熱，再調至中慢火炒香配料；上碟時建議用厚瓦碟盛着瓦鍋，小心燙傷。

What should be noted when cooking with clay pot?
Preheat clay pot over low heat for a while then stir-fry the condiments over medium-low heat until fragrant. Also it is recommended to place thick clay dish below the clay pot on the dining table.

泡菜年糕牛仔骨

Beef Short Ribs with Kimchi and Rice Cakes

◎ 材料（4 人份量）

牛仔骨 300 克

韓國泡菜半杯

洋蔥 1/4 個（切幼絲）

甜青椒半個（去籽，切幼絲）

紅蘿蔔 1/4 個（切幼絲）

年糕 100 克

韓式辣椒醬 1.5 湯匙

蒜茸 1 茶匙

◎ 醃料

鹽半茶匙

黑椒粉及粟粉各少許

水 1 湯匙

◎ 調味料

鹽半茶匙

生抽 1 茶匙

泡菜汁 1 湯匙

麻油少許

◎ 做法

1. 牛仔骨解凍，於兩節骨之間切成塊狀，用刀背拍鬆，加入醃料拌勻。

2. 燒熱油 1 湯匙，下牛仔骨煎至兩面七成熟，盛起。

3. 燒熱少許油，下甜青椒絲略炒，盛起。

4. 燒熱少許油，加入洋蔥、蒜茸及紅蘿蔔絲略炒，鋪上年糕片，傾入水 1/4 杯加蓋煮 5 分鐘，至年糕變軟。

5. 加入調味料及韓式辣醬炒勻，下泡菜及青椒絲炒勻，最後牛仔骨回鑊快炒即可。

◎ Ingredients (Serves 4)

300 g beef short ribs

1/2 cup Korean kimchi

1/4 onion (finely shredded)

1/2 green bell pepper (deseeded; finely shredded)

1/4 carrot (finely shredded)

100 g rice cakes

1.5 tbsp Korean chilli sauce

1 tsp finely chopped garlic

◎ Marinade

1/2 tsp salt

ground black pepper

cornflour

1 tbsp water

◎ Seasoning

1/2 tsp salt

1 tsp light soy sauce

1 tbsp kimchi sauce

sesame oil

◎ Method

1. Defrost the beef short ribs. Cut between the bones into pieces. Tenderize by pounding with the back of a knife. Mix with the marinade.

2. Heat up 1 tbsp of oil. Fry the beef short ribs until both sides are 70% done. Remove.

3. Heat up a little oil. Slightly stir-fry the green bell pepper. Set aside.

4. Heat up a little oil. Slightly stir-fry the onion, garlic and carrot. Lay the rice cakes on top. Pour in 1/4 cup of water. Put a lid on. Cook for 5 minutes until the rice cakes turn soft.

5. Put in the seasoning and Korean chilli sauce. Stir-fry evenly. Mix in the kimchi and bell pepper. Add the beef short ribs and stir-fry quickly to finish.

◎◎ 零失敗技巧 ◎◎
Successful Cooking Skills

買不到韓式辣椒醬，怎辦？
可用茄汁及少許辣椒粉拌勻使用；但香味始終不及韓式辣椒醬。
What to do if Korean chilli sauce is not available?
Mix ketchup with a little paprika instead, but it is less fragrant than the Korean chilli sauce.

甚麼是泡菜汁？
即泡菜內之汁料，酸酸辣辣，香味濃郁！
What is kimchi sauce?
That is the sauce in the kimchi. It tastes sour and spicy, and smells strong!

炒年糕容易黏底，怎辦？
先將蔬菜鋪於年糕底，加少許水焗煮，待年糕變軟快炒即可。
How to avoid rice cakes sticking to the pan?
Lay some vegetables under the rice cakes. Add a little water and cook until the rice cakes turn soft. Then stir-fry quickly.

蒜香豆豉辣椒醬香草炒蜆

Stir-fried Clams with Mint, Black Bean and Garlic Chilli Sauce

◎ 材料 （4 人份量）

活蜆 1 斤
青甜椒 1 個
紅甜椒 1 個
蒜茸 1 湯匙
薄荷葉 2 棵
豆豉蒜香辣椒醬 2 茶匙（見 p.86）

◎ 調味料

蠔油 2 湯匙
麻油 1 茶匙

◎ 獻汁（拌勻）

粟粉 1 茶匙
水 2 湯匙

◎ 做法

1. 活蜆用清水浸半天（加入不鏽鋼茶匙，有助吐淨砂粒），洗淨。

2. 青、紅甜椒去蒂、去籽，洗淨，切塊。

3. 活蜆放入滾水灼 1 分鐘，盛起，隔去水分。

4. 燒熱鑊下油 2 湯匙，下辣椒醬、蒜茸及青紅甜椒炒香，加入蜆，灒酒炒勻，下調味料及熱水半杯，煮片刻至蜆張開及全熟，下薄荷葉，最後埋獻即成。

◎ Ingredients (Serves 4)

600 g live clams
1 green bell pepper
1 red bell pepper
1 tbsp finely chopped garlic
2 sprigs mint (leaves only)
2 tsp black bean and garlic chilli sauce (see p.86)

◎ Seasoning

2 tbsp oyster sauce
1 tsp sesame oil

◎ Thickening glaze (mixed well)

1 tsp cornflour
2 tbsp water

◎ Method

1. Soak the live clams in fresh water with stainless steel teaspoon for 1/2 day so that they spit out the sand. Rinse well.

2. Cut of the stems of the bell peppers. Seed them and rinse. Cut into pieces.

3. Blanch the clams in boiling water for 1 minute. Drain.

4. Heat a wok and add 2 tbsp of oil. Put in the chilli sauce, garlic and bell peppers. Stir fry until fragrant. Then put in the clams and sizzle with wine. Stir well. Add seasoning and 1/2 cup of hot water. Cook briefly until all clams are open. Add the mint. Stir in thickening glaze and bring to the boil again. Serve.

* 蒜香豆豉辣椒醬
Black Bean and Garlic Chilli Sauce

◎ 材料
指天椒 3 兩
蒜茸 2 湯匙
豆豉 2 湯匙
蝦米 2 湯匙
粟米油 1.5 杯

◎ 調味料
老抽 1 湯匙
鹽 1 茶匙
糖 1.5 茶匙

◎ 做法
1. 指天椒洗淨，去蒂，切碎；蝦米洗淨，切碎；豆豉用水沖洗，切碎（或用石舂將材料舂爛）。
2. 下油半杯燒熱，下蝦米及指天椒炒香，加入蒜茸及豆豉茸不斷炒香，下調味料及餘下之粟米油，用小火煮至滾，再煮片刻，待涼，入瓶儲存。

◎ Ingredients
113 g bird's eye chillies
2 tbsp finely chopped garlic
2 tbsp fermented black beans
2 tbsp dried shrimps
1.5 cups corn oil

◎ Seasoning
1 tbsp dark soy sauce
1 tsp salt
1.5 tsp sugar

◎ Method
1. Rinse the bird's eye chillies. Remove the stems. Finely chop them. Set aside. Rinse the dried shrimps. Chop them. Set aside. Rinse the fermented black beans. Chop or crush them. (Or you may crush ingredients separately with a mortar and pestle.)
2. Heat a wok and pour in 1/2 cup of oil. Stir fry dried shrimps and bird's eye chillies until fragrant. Add garlic and fermented black bean. Stir continuously until fragrant. Add seasoning and the remaining corn oil. Bring to the boil over low heat. Cook briefly. Leave it to cool. Transfer into sterilized bottles.

◯◯ 零失敗技巧 ◯◯
Successful Cooking Skills

蜆先燙再炒，肉質欠鮮味嗎？

不會。只是將蜆略燙一下，蜆肉仍然飽滿，下鍋快炒容易熟透，而且可挑出沒張開的壞死蜆。

You blanch the clams before stir-frying them. Will their flavour be lost in the blanching water?

No, it won't. You just blanch them very briefly so that the clams are undercooked before stir-frying. This step speeds up the stir-frying time and you can pick out those dead ones (those not opening) and discard them.

炒煮蒜香豆豉辣椒醬有何要訣？

加入蒜茸及豆豉茸後，必須使用小火炒，切勿炒焦而影響味道。

What is the secret trick of making the black bean and garlic chilli sauce?

Stir fry the mixture over low heat after adding finely chopped garlic and crushed fermented black beans. Otherwise it might get burnt and turns bitter.

蒜香豆豉辣椒醬與蜆肉如何搭配？

蒜香、豆豉濃、紅椒辣，最能突出海鮮食材的鮮味。

How does the black bean and garlic chilli sauce work with clams?

The strong garlicky taste, the bean flavour and piquancy of the chillies accentuate the seafood flavour of the clams.

黑椒醬爆蟶子皇

Stir-fried King Razor Clams with Black Pepper Sauce

◎ 材料 （4 人份量）
蟶子皇 8 隻（約 1.5 斤）
指天椒 2 隻（切圈）
蒜茸 1 湯匙
薑米 2 茶匙
黑胡椒碎 4 茶匙
葱絲 1 湯匙

◎ 調味料
水 4 湯匙
老抽 4 茶匙
鹽 1 茶匙
糖半茶匙
胡椒粉少許
生粉 2.5 茶匙

◎ 做法
1. 蟶子皇的薄膜剝開，去腸臟（售者可代勞），洗淨。
2. 煮滾適量清水（以浸過蟶子為宜），下蟶子略灼半分鐘，盛起，瀝乾水分。
3. 燒熱油爆香黑胡椒碎、紅椒圈、蒜茸及薑米，注入調味料煮滾，蟶子回鑊快速炒勻，上碟，灑上葱絲裝飾即成。

Ingredients (Serves 4)

8 king razor clams (about 900 g)
2 bird's eye chilies (cut into rings)
1 tbsp finely chopped garlic
2 tsp chopped ginger
4 tsp chopped black pepper
1 tbsp shredded spring onion

Seasoning

4 tbsp water
4 tsp dark soy sauce
1 tsp salt
1/2 tsp sugar
ground white pepper
2.5 tsp caltrop starch

Method

1. Cut open thin membrane of king razor clams and remove the entrails (or ask the monger for help). Rinse.

2. Bring water to the boil (to cover all razor clams). Blanch razor clams briefly for about 30 seconds. Drain.

3. Heat oil in wok. Stir-fry chopped black pepper, chilli rings, finely chopped garlic and chopped ginger until fragrant. Pour in seasoning and bring to the boil. Add the razor clams and stir-fry quickly. Put into a plate. Sprinkle over shredded spring onion for garnishing. Serve.

零失敗技巧
Successful Cooking Skills

如何挑選蟶子皇？
選鮮活及體型大的蟶子為佳，肉多汁豐。
How to choose king razor clams?
Choose fresh and large king razor clams that are fleshy and juicy.

蟶子肉容易炒至過韌嗎？
蟶子肉容易炒至韌，建議快手拌炒上碟。
Would razor clams stir-fried to hard texture easily?
Yes. You are suggested to stir-fry them quickly and serve.

麻辣炮椒雞鍋

Spicy Bullet Chilli Chicken Casserole

◯◯ **材料 （3 至 4 人份量）**

光雞 12 兩
四川花椒 2 茶匙
炮彈辣椒半兩
薑絲 1 湯匙
葱段 1 條
清雞湯 1/4 杯

◯◯ **醃料**

鹽 1/4 茶匙
生抽 2 茶匙
老抽 1 茶匙
糖 1/8 茶匙
胡椒粉少許
生粉半湯匙
油半湯匙

◯◯ **做法**

1. 光雞洗淨，抹乾水分，斬件，加入醃料拌勻醃 15 分鐘

2. 熱鑊下油，下薑絲及雞件爆香，拌炒約 5 分鐘至雞件半熟，盛起。

3. 瓦鍋下油，爆香花椒及炮椒，下雞件爆炒，傾入清雞湯，加蓋，用慢火煮至汁液收乾（約 6 分鐘），加入葱段多煮片刻，原鍋上桌。

Ingredients (Serves 3-4)

450 g chicken
2 tsp Sichuan peppercorns
19 g dried bullet chillies
1 tbsp shredded ginger
1 sprig sectioned spring onion
1/4 cup chicken stock

Marinade

1/4 tsp salt
2 tsp light soy sauce
1 tsp dark soy sauce
1/8 tsp sugar
ground white pepper
1/2 tbsp caltrop starch
1/2 tbsp oil

Method

1. Rinse the chicken. Wipe dry. Chop into pieces. Mix with the marinade and rest for 15 minutes.

2. Add oil in a heated wok. Fry the ginger and chicken until fragrant. Stir-fry for about 5 minutes until the chicken is half done. Set aside.

3. Add oil in a casserole. Stir-fry the Sichuan peppercorns and bullet chillies until fragrant. Put in the chicken and stir-fry. Pour in the chicken stock. Cover with the lid. Simmer until the sauce dries (about 6 minutes). Add the spring onion and cook for a while. Serve with the casserole.

零失敗技巧
Successful Cooking Skills

雞鍋的食味如何？
麻、香、辣，嗜辣者絕對享受味覺上的震憾！
How does the chicken casserole taste?
Tingly-numbing, fragrant, hot – a stunning sensation to those loving spicy food!

四川出產的花椒有何特色？
紅彤彤的四川花椒，香麻味濃，品質較佳，香料店舖有售。
What are the characteristics of Sichuan peppercorns?
Sichuan peppercorns are red, tingly-numbing, fragrant and better in quality. They can be found in spice shops.

甚麼是炮彈辣椒？
炮彈辣椒又名雞心椒或彈子椒，外型圓小，辣味濃，但比指天椒溫和，香料店舖有售。
What is bullet chilli?
It is small, round and pungent but less spicy compared with bird's eye chilli. It is available at spice shops.

XO醬翠玉瓜炒魚塊

Stir-fried Fish Fillet with Zucchini in XO Sauce

材料 （4 人份量）
冰鮮石斑肉 6 兩
翠玉瓜 1 個
蒜肉 2 粒
薑 4 片
XO 醬 2 茶匙（見 p.94）
紹酒半湯匙

醃料
胡椒粉少許
蛋白半個
粟粉 1 茶匙

調味料
蠔油 1 湯匙
粟粉半茶匙
水 2 湯匙

做法
1. 石斑肉洗淨，切塊，下醃料拌勻，備用。
2. 翠玉瓜洗淨，開邊、去籽，切斜塊。
3. 石斑魚塊放入油鑊內，煎至微黃色及八成熟，盛起。
4. 燒熱鑊下油 2 湯匙，下薑片及蒜肉炒香，下翠玉瓜炒勻，加入 XO 醬及石斑魚塊，濽酒炒勻，最後加入調味料炒片刻，至汁液濃稠即成。

Ingredients (Serves 4)
225 g chilled garoupa fillet
1 zucchini
2 cloves skinned garlic
4 slices ginger
2 tsp XO sauce (see p.94)
1/2 tbsp Shaoxing wine

Marinade
ground white pepper
1/2 egg white
1 tsp cornflour

Seasoning
1 tbsp oyster sauce
1/2 tsp cornflour
2 tbsp water

Method
1. Rinse the garoupa fillet. Cut into pieces. Add marinade and mix well. Set aside.
2. Rinse the zucchini. Cut into halves and scoop out the seeds. Cut into piece at an angle.
3. Fry the garoupa pieces in some oil in a wok until lightly browned on both sides and medium-well done. Set aside.
4. In the same wok, heat up 2 tbsp of oil. Stir fry ginger and garlic cloves until fragrant. Put in the zucchini and stir well. Add XO sauce and garoupa pieces. Sizzle with wine. Stir well. Lastly add seasoning and stir briefly. Cook until the sauce reduces. Serve.

*XO 醬
**XO Sauce*

◎ 材料

乾瑤柱 1 兩
蝦乾 1 兩
金華火腿茸 1/4 兩
指天椒 1.5 兩
蒜茸 1 湯匙
乾葱茸半湯匙
粟米油 1.5 杯
米酒 1 湯匙

◎ 調味料

蠔油 1 湯匙
鹽 1 茶匙
糖半茶匙

◎ 做法

1. 乾瑤柱及蝦乾用水浸軟，瑤柱壓成絲；蝦乾切碎。

2. 指天椒洗淨，去蒂，切粒。

3. 燒熱半杯油，下乾瑤柱絲及蝦乾炒香，炒約 10 分鐘至起泡，潷酒，下乾葱茸、指天椒、蒜茸及金華火腿茸炒香，下調味料炒勻，注入餘下之油分，煮至油滾，再煮片刻即成（撇去油面之泡沫），待涼，入瓶儲存。

◎ Ingredients

38 g dried scallops
38 g dried prawns
10 g grated Jinhua ham
57 g bird's eye chillies
1 tbsp finely chopped garlic
1/2 tbsp finely chopped shallot
1.5 cups corn oil
1 tbsp rice wine

◎ Seasoning

1 tbsp oyster sauce
1 tsp salt
1/2 tsp sugar

◎ Method

1. Soak the dried scallops and dried prawns in water until soft. Tear the dried scallops into shreds. Finely chop the dried prawns.

2. Rinse the bird's eye chillies. Cut off the stems and dice them.

3. Heat a wok and add 1/2 cup of oil. Stir fry dried scallops and dried prawns until fragrant. Keep on stirring for about 10 minutes until it bubbles. Sizzle with wine. Add shallot, bird's eye chillies, garlic and Jinhua ham. Stir fry until fragrant. Add seasoning and stir well. Pour in the remaining oil. Cook until the oil boils. Keep on cooking a bit longer. Skim off the foam on the surface. Leave it to cool. Store in sterilized bottles.

◎ 零失敗技巧 ◎
Successful Cooking Skills

如何避免自家製的醬不發霉？

瑤柱及蝦乾含水分，必須炒至乾透，讓水分充份揮發，不容易發霉。

What is the technique for avoiding the homemade sauce get mouldy easily?

Dried scallops and dried prawns contain water. That's why they should be stir-fried until dry to remove its moisture content. That will ensure the XO sauce will not go stale or mouldy easily.

為甚麼不先爆香 XO 醬？

自家炮製的 XO 醬，甘香味特濃，故毋須預先爆香。

Why don't you fry the XO sauce first?

The homemade XO sauce has been fried until fragrant and dry previously. Thus, it is fragrant enough as it is and you don't need to fry it again.

石斑魚塊為何加蛋白醃製？

令石斑魚塊的肉質嫩滑，口感佳！

Why do you marinate the fish with egg white?

The egg white coats the fish and keeps it tender and juicy.

椰香咖喱蟹
Curry Crab with Coconut Milk

◎ 材料（4 人份量）
肉蟹 1 隻（約 1 斤重）
咖喱醬 1.5 湯匙
椰漿 125 毫升
紅辣椒 1 隻
乾葱頭 1 粒
蒜頭 2 粒
薑 2 片

◎ 調味料
清雞湯 125 毫升
鹽 1/3 茶匙
魚露 1 湯匙
糖 1/4 茶匙
胡椒粉少許
清水 60 毫升

◎ 做法
1. 肉蟹洗淨，斬件，瀝乾水分。
2. 咖喱醬與水 1 湯匙拌勻。
3. 紅辣椒、乾葱頭及蒜頭切碎。
4. 燒熱適量油，下蟹件炒至轉成紅色，盛起。
5. 熱鑊下油，下辣椒、乾葱頭、蒜頭及薑片爆香，放入咖喱醬用小火爆香，放入蟹件及調味料拌勻，加蓋煮 10 分鐘。
6. 最後傾入椰漿煮滾，上碟即成。

Ingredients (Serves 4)

1 mud crab (about 600 g)
1.5 tbsp curry sauce
125 ml coconut milk
1 red chilli
1 shallot
2 cloves garlic
2 slices ginger

Seasoning

125 ml chicken broth
1/3 tsp salt
1 tbsp fish sauce
1/4 tsp sugar
ground white pepper
60 ml water

Method

1. Rinse mud crab. Chop into pieces and drain.

2. Mix curry sauce with 1 tbsp of water.

3. Chop red chilli, shallot and garlic.

4. Heat oil in a wok. Stir-fry crab until turns red. Drain.

5. Add oil into a hot wok. Stir-fry red chilli, shallot, garlic and ginger slices until fragrant. Put in curry sauce and stir-fry quickly over low heat until fragrant. Put in the crab and seasoning. Mix well and cover the lid. Cook for about 10 minutes.

6. Lastly pour in coconut milk and bring to the boil. Serve.

零失敗技巧
Successful Cooking Skills

咖喱醬為何用小火炒煮？
令咖喱香氣慢慢散發出來，而且避免咖喱醬焦燶，影響食味。
Why stir-frying curry sauce over low heat?
This makes the curry fragrance emit out slowly and also avoids it get charred.

如何保持濃郁的椰香味？
最後傾入椰漿，勿煮太久，上桌時椰香四溢。
How to keep the rich coconut smell?
Pour in the coconut milk at the last and do not cook for too long, there is still rich coconut smell when serving.

香茅辣椒炒東風螺

Stir-fried Spiral Babylon with Lemongrass and Chillies

◎ 材料 （4 人份量）
東風螺 1 斤
香茅 2 枝
指天椒 3 至 4 隻
蒜頭 1 粒

◎ 調味料
清水 2/3 杯
鹽 2/3 茶匙
糖半茶匙
老抽半湯匙
魚露 3/4 湯匙
生粉 1 茶匙

◎ 做法

1. 東風螺洗淨，用淡鹽水浸 1 小時或以上（有助吐沙），烹調前再洗一次。

2. 香茅切斜段；指天椒切小片；蒜頭切片。

3. 煮滾一鍋水，放入東風螺飛水 2 分鐘，盛起。

4. 燒熱適量油，下蒜片、香茅及辣椒炒香，注入調味料煮滾片刻，加入東風螺拌炒至汁液濃稠，上碟享用。

⦿ Ingredients (Serves 4)

600 g spiral babylon
2 stalks lemongrass
3 to 4 bird's eye chillies
1 clove garlic

⦿ Seasoning

2/3 cup water
2/3 tsp salt
1/2 tsp sugar
1/2 tbsp dark soy sauce
3/4 tbsp fish sauce
1 tsp caltrop starch

⦿ Method

1. Rinse spiral babylon. Soak in diluted salted water for 1 hour or above to let them spit out sand. Rinse once again before cooking.

2. Section lemongrass at an angle. Cut bird's eye chillies into small slices. Slice garlic.

3. Bring a pot of water to the boil. Scald spiral babylon for 2 minutes and drain.

4. Heat oil in a wok. Stir-fry garlic slices, lemongrass and chillies until fragrant. Pour in the seasoning and boil for a while. Add spiral babylon and stir-fry until the sauce thickens. Serve.

⦿ 零失敗技巧 ⦿
Successful Cooking Skills

惹味的秘訣在於哪個步驟？
重點在於爆香蒜片、香茅及辣椒，令香氣帶入東風螺肉。
Which step may I need to pay attention for this savoury dish?
Should be stir-fry the garlic, lemongrass and red chilli first, the aroma will be add to the spiral babylon.

為甚麼東風螺先飛水才炒煮？
令附在螺身的雜質掉落水中，去掉雜質，食用時更健康。
Why scald the spiral babylon before stir-frying?
This can remove any impurities in the spiral babylon.

此菜可以作為頭盤冷吃嗎？
絕對可以，建議可預早一天烹調，適當冷藏，食味更佳。
Can this dish become a cold appetite?
Absolutely. It is suggested to cook a day in advance and store in the refrigerator.

魚腸涼瓜煎蛋

Egg Omelette with
Fish Intestines and Bitter Melon

材料 （4 人份量）

新鮮鯇魚腸 2 副
苦瓜 6 兩
雞蛋 4 個
鹹菠蘿 1 塊（見 p.102）
糖 1 茶匙

做法

1. 用剪刀剪開鯇魚腸，撕去魚脂肪，只取魚腸及魚肝，用鹽擦淨，沖洗，浸於米醋水待片刻，飛水，瀝乾水分。

2. 涼瓜開邊、去籽，洗淨，切薄片，飛水，過冷河，擠乾水分。

3. 鹹菠蘿切碎；雞蛋拂勻。

4. 蛋漿、鹹菠蘿、鯇魚腸、魚肝及涼瓜拌勻，傾入油鑊用中小火煎成凝固的厚蛋餅，再煎至兩面金黃色，趁熱享用。

Ingredients (Serves 4)

2 sets of freshly slaughtered grass carp intestines
225 g bitter melon
4 eggs
1 chunk salted pineapple (see p.102)
1 tsp sugar

Method:

1. Cut open the fish intestines along the length with a pair of scissors. Trim off the fat. Use only the fish intestines and liver. Rub salt all over. Rinse well. Soak them in diluted rice vinegar briefly. Blanch them in boiling water. Drain.

2. Cut the bitter melon into halves. Seed it and rinse well. Slice thinly. Blanch in boiling water and rinse in cold water. Squeeze dry.

3. Finely chop the salted pineapple. Whisk the eggs well.

4. In a large mixing bowl, put in the whisked eggs, salted pineapple, fish intestines, fish liver and bitter melon. Mix well. Pour into a greased wok. Fry over medium-low heat until set. Flip the omelette over to fry the other side. Fry until both sides golden. Serve hot.

＊鹹菠蘿製法
*Making salted pineapple

魚腸涼瓜煎蛋

◯◯ 材料
新鮮菠蘿 1 個
潮州豆醬 1 瓶（大）
玻璃瓶 1 個

◯◯ Ingredients:
1 fresh pineapple
1 large bottle Chaozhou salted bean sauce
1 large glass container

◯◯ 做法
1. 菠蘿削去皮，洗淨，抹乾水分，切塊。
2. 將菠蘿塊放於玻璃瓶內，加入潮州豆醬（浸過菠蘿面少許），加蓋密封，冷藏 5 至 7 日即可。

◯◯ Method:
1. Peel the pineapple and rinse well. Wipe dry and cut into chunks.
2. Arrange the pineapple chunks in the glass container. Pour in the Chaozhou salted bean sauce to cover the pineapple. Cover the lid and seal well. Refrigerate for 5 to 7 days.

◎◎ 零失敗技巧 ◎◎
Successful Cooking Skills

製作鹹菠蘿有何秘訣？

新鮮菠蘿含水量多，必須用廚房紙吸乾水分，以免製成醬料後味道變淡及變壞，影響質素。

What is the tip for making salted pineapple?

Fresh pineapple has high water content. You have to wipe it completely dry with paper towel. Otherwise, the salted pineapple will taste bland and turn stale easily.

必須購買鯇魚腸？帶腥味嗎？

由於鯇魚肥美，故其腸臟較大。經鹽洗擦及用米醋水浸透後，魚腸及魚肝不帶魚腥味。

Must I use grass carp intestines? Do they taste fishy?

Because of their size and fattiness, grass carps usually have thicker intestines. That's why they are preferred. After rubbing salt on the intestines, they are further soaked in diluted rice vinegar. Such steps help remove the fishy taste from the intestines and liver.

魚販可代清洗魚腸嗎？

相熟的魚販可以。若自行清洗，緊記別弄破魚膽，以免魚腸沾滿苦澀汁液。

Can I ask the fishmonger to clean the fish intestines for me?

You may talk your favourite fishmonger into cleaning them for you, if you have a good relation with him/her. If you're cleaning them yourself, make sure you don't break the gall in the process. Otherwise, the fish intestines will taste bitter.

香酥芝麻青芥辣雞中翼

Deep-fried Chicken Wings
with Sesame Seeds and Wasabi

材料（3 至 4 人份量）

雞中翼 8 隻（約 8 至 10 兩）
日式青芥辣 3/4 湯匙
蛋黃醬 3 湯匙
白芝麻 1.5 湯匙（炒香）
凍開水 1.5 湯匙

醃料

鹽 1/3 茶匙
魚露半湯匙
雞粉半茶匙
糖 3/4 茶匙
生粉半湯匙

脆漿料

麵粉 4 湯匙
發粉 1/3 茶匙
生粉 1 湯匙
清水 5 湯匙
油半湯匙（後下）

做法

1. 雞翼解凍，洗淨，抹乾水分，下醃料拌勻醃 15 分鐘。

2. 脆漿料拌勻，待半小時後，下油拌勻，備用。

3. 蛋黃醬、青芥辣及凍開水拌勻，備用。

4. 熱鑊下油，雞翼沾上脆漿料，放入熱油炸熟，瀝乾油分。

5. 用慢火煮滾步驟（3）的醬料，放入雞翼拌勻醬料，上碟，最後灑上芝麻即可。

Ingredients (Serves 3-4)

8 chicken mid-joint wings (about 300 g to 375 g)
3/4 tbsp wasabi
3 tbsp mayonnaise
1.5 tbsp white sesame seeds (toasted)
1.5 tbsp cold drinking water

Marinade

1/3 tsp salt
1/2 tbsp fish sauce
1/2 tsp chicken bouillon powder
3/4 tsp sugar
1/2 tbsp caltrop starch

Batter ingredients

4 tbsp flour
1/3 tsp baking powder
1 tbsp caltrop starch
5 tbsp water
1/2 tbsp oil (added at last)

Method

1. Defrost the chicken wings. Rinse and wipe them dry. Mix with the marinade and rest for 15 minutes.

2. Mix the batter ingredients well. Rest for 1/2 hour. Mix in the oil. Set aside.

3. Mix the mayonnaise, wasabi and cold drinking water evenly. Set aside.

4. Add oil in a heated wok. Dip the chicken wings in the batter. Deep-fry in the hot oil until done. Drain.

5. Bring the sauce from step (3) to the boil over low heat. Put in the chicken wings and mix evenly. Arrange on a plate. Sprinkle the sesame seeds on top to serve.

◎ 零失敗技巧 ◎
Successful Cooking Skills

如何炸成金黃香脆的雞翼？

除了脆漿料調得妥當外，建議使用新油炸雞翼，但家用的油量有限，別一次過放太多雞翼入鍋，以免油溫瞬間下降，影響效果。

How to make deep-fried chicken wings golden and crispy?

Apart from making a proper batter, use fresh oil for deep-frying. We will not use a lot of oil at home, and so do not put in too many chicken wings at a time. Otherwise, the temperature of oil will drop making the deep-frying process less effective.

如何有效解凍雞翼？

放入雪櫃的下層自然解凍，當然是最佳的方法；若時間不多，將雞翼放於食物袋密封，用水沖 20 分鐘。

How to defrost the chicken wings effectively?

The best way is to put them in the lower shelf of the refrigerator to let them defrost naturally. If you are in a rush, put them in a zipper bag and rinse under tap water for 20 minutes.

脆炸門鱔肉

Deep-fried Conger-pike Eel

脆
炸
門
鱔
肉

◎ 材料 （4 至 5 人份量）
門鱔魚 1 斤

◎ 醃料
胡椒粉少許
幼海鹽半茶匙
粟粉 2 茶匙（後下）

◎ 脆漿料
自發粉半杯
水 1/3 杯
油 2 湯匙
鹽 1 茶匙
胡椒粉少許
＊ 調勻，待半小時

◎ 做法
1. 門鱔魚洗淨，起肉及切塊，魚骨切塊留用。
2. 門鱔肉與醃料拌勻，醃半小時，再加入粟粉拌勻。
3. 魚塊放入脆漿料內拌勻，逐塊放入滾油炸至金黃及全熟，隔油，上碟。

◎ Ingredients (Serves 4-5)
600 g conger-pike eel

◎ Marinade
ground white pepper
1/2 tsp fine sea salt
2 tsp cornflour (added at last)

◎ Batter
1/2 cup self-raising flour
1/3 cup water
2 tbsp oil
1 tsp salt
ground white pepper
* mixed well and leave for 1/2 hour

◎ Method
1. Rinse the conger-pike eel. Remove the meat from the bone. Cut the meat and bone into pieces. Reserve the bones.
2. Mix the meat with the marinade and rest for 1/2 hour. Add the cornflour and mix well.
3. Put the meat into the batter. Mix well. Put each piece of the meat into boiling oil. Deep-fry until golden and fully cooked. Drain and serve.

◎ 零失敗技巧 ◎
Successful Cooking Skills

炸門鱔肉的食味如何？
門鱔肉質厚，鮮味是魚類之冠。灑少許醃料快速炸透，香脆又美味！
What does conger-pike eel taste like?
Prized for its rich flavour, it is fleshy and possible one of the most tasty fish. Just sprinkle seasoning on top and fry over high heat quickly until done. It's crispy and delicious.

魚肉要裹上大量脆漿嗎？
只要均勻地輕輕裹上脆漿即可，別吃下厚厚的脆漿粉糰啊！
Need to coat the meat with a lot of batter for deep-frying?
Only a thin and even layer will do. Do not try a thick batter!

魚肉大約炸多久才熟透？
建議炸約 5 分鐘，魚肉必定熟透，太久容易令魚肉粗韌。
Deep-frying for how long to make it done?
It should be perfectly done to be deep-fried for about 5 minutes. Long cooking makes the meat tough.

南乳碎炸雞
Deep-fried Chicken with Fermented Tarocurd

◎ 材料（4 人份量）

光雞 12 兩（約半隻）
南乳 1.5 湯匙
葱絲少許
生粉 4 茶匙

◎ 醃料

鹽 1/3 茶匙
沙薑粉半茶匙
五香粉 1/3 茶匙
玫瑰露酒半湯匙
蛋液 1.5 湯匙

◎ 做法

1. 光雞洗淨，斬件，抹乾水分，加入醃料及南乳拌勻醃半小時。
2. 雞件灑上生粉，拌勻。
3. 燒熱適量油，放入雞件炸熟，上碟，以葱絲伴碟，趁熱享用。

Ingredients (Serves 4)

450 g chicken (about 1/2 chicken)
1.5 tbsp fermented tarocurd
shredded spring onion
4 tsp caltrop starch

Marinade

1/3 tsp salt
1/2 tsp spice ginger powder
1/3 tsp five-spice powder
1/2 tbsp Chinese rose wine
1.5 tbsp egg wash

Method

1. Rinse the chicken. Chop into pieces. Wipe them dry. Mix with the marinade and fermented tarocurd. Rest for 1/2 hour.

2. Sprinkle the chicken with the caltrop starch. Mix well.

3. Heat some oil. Deep-fry the chicken until fully done. Put on a plate. Garnish with spring onion. Serve.

零失敗技巧
Successful Cooking Skills

雞塊用南乳等調味料拌勻，味道會太鹹嗎？
由於南乳的用量不多，故味道不太鹹，而且炸雞塊的味道尤如南乳吊燒雞，但製法簡便得多！

Will it be too salty by mixing the chicken pieces with the fermented tarocurd and other seasoning?

It will not be too salty as a small amount of fermented tarocurd is used. The flavour of the dish is similar to that of Roasted Chicken with Fermented Tarourd, but the cooking method is easier.

怎樣將雞件均勻地沾上生粉？
教你一個方法：將生粉及雞件放入食物袋內，拉緊袋口，上下搖動，能均勻地沾上生粉。

How to coat the chicken with caltrop starch evenly?

Here is a method: Put the caltrop starch and chicken into a food storage bag. Hold the opening of the bag tightly. Shake the bag up and down.

炸雞的火候怎樣調校？
建議火候勿太猛，否則雞件的顏色容易轉深，但雞肉仍未熟透。

How to adjust the heat of deep-frying chicken?

The heat for deep-frying should not be too high for easy control of its doneness; otherwise, the colour of the chicken will turn deep while the inside is still raw.

零失敗
秘方系列

煮出香濃
伴飯餸
Rich and delectable home dishes

編者
Forms Kitchen編輯委員會

Editor
Editorial Committee, Forms Kitchen

美術設計
馮景蕊

Design
Carol Fung

排版
劉葉青

Typography
Rosemary Liu

出版者

香港鰂魚涌英皇道1065號
東達中心1305室
電話
傳真
電郵
網址

Publisher
Forms Kitchen
Room 1305, Eastern Centre, 1065 King's Road,
Quarry Bay, Hong Kong.
Tel: 2564 7511
Fax: 2565 5539
Email: info@wanlibk.com
Web Site: http://www.wanlibk.com
            http://www.facebook.com/wanlibk

發行者
香港聯合書刊物流有限公司
香港新界大埔汀麗路36號
中華商務印刷大廈3字樓
電話
傳真
電郵

Distributor
SUP Publishing Logistics (HK) Ltd.
3/F., C&C Building, 36 Ting Lai Road,
Tai Po, N.T., Hong Kong
Tel: 2150 2100
Fax: 2407 3062
Email: info@suplogistics.com.hk

承印者
中華商務彩色印刷有限公司

Printer
C & C Offset Printing Co., Ltd.

出版日期
二零一九年五月第一次印刷

Publishing Date
First print in May 2019

鳴謝以下作者提供食譜（排名不分先後）：
黃美鳳、Feliz Chan、Winnie姐